DICTIONARY OF AGRICULTURE

DICTIONARY OF
AGRICULTURE

Gurdeep Chhatwal
M.Sc., Ph.D., M.I.A., H.P.S.

ANMOL PUBLICATIONS PVT. LTD.
NEW DELHI - 110 002 (INDIA)

ANMOL PUBLICATIONS PVT. LTD.
Regd. Office: 4360/4, Ansari Road, Daryaganj,
New Delhi-110002 (India)
Tel.: 23278000, 23261597, 23286875, 23255577
Fax: 91-11-23280289
Email: anmolpub@gmail.com
Visit us at: www.anmolpublications.com

Branch Office: No. 1015, Ist Main Road, BSK IIIrd Stage
IIIrd Phase, IIIrd Block, Bangalore-560 085 (India)
Tel.: 080-41723429 • Fax: 080-26723604
Email: anmolpublicationsbangalore@gmail.com

Dictionary of Agriculture

First Edition 1998

Reprint 1989, 1990, 1991, 1992, 1993, 1994, 1995

Second Edition 1996

Reprint 2003, 2006, 2008, 2010

PRINTED IN INDIA

Printed at Atul Printers, Delhi.

Preface

It is a matter of great satisfaction that the first edition of this dictionary got exhausted within a short period. Now I am presenting the new revised and enlarged second edition. This has been revised to make it uptodate

This dictionary has been written to provide a reference source for students of agriculture, farmers and all those involved in or associated with the agriculture industry in whatever capacity. It is concerned essentially with agriculture and embraces such horticultural botanical, geographical, biological, physical, geomorphological and chemical terms as may be needed to provide adequate information. Entries also include the most common and agriculturally significant diseases and pests of farm animals or crops with their symptoms.

The terms included in this dictionary have been presented in brief analytical phrases thereby avoiding the more comprehensive type of treatment appropriate to large reference works. The extent of each entry has been determined by the need for definition of meaning, function and agriculture relevance. In this dictionary entries have been added to take account of recent developments and trends in agricultural industry.

The editor is happy to give credit here to all those who had a hand in getting the dictionary into its present form and particularly to put on record my great debt to Dr. M.S. Yadav for the care and interest he took in clearing up complex points so that they might be put in simple English.

In compiling a dictionary of this kind it becomes necessary to draw upon the work of many authorities and seek the advice of colleagues to all of whom the editor is deeply indebted.

Editor

Preface

A

A. Ratio designating relative thickness or thinness of nematode. It is found out by dividing length of nematode body by its maximum width.

Abaction. The stealing of cattle.

ABC Soil. Soil having a distinctly developed profiles including A, B and C horizons.

Aberdeen-Angus. A breed of black, hornless beef cattle, developed using interbreeding from a mixture of breeds. They have a long, deep body.

A Bullock Work Unit. Refers to an average amount of work which has been put up by one bullock in eight hours.

AC Soil. Soil which has a profile containing only A and C horizons but no clearly developed B horizon.

A Chromosome. Chromosome belonging to normal chromosome complement.

A Horizon. The term used for the surface and sub-surface soil which is having most of the organic matter and is subject to leaching.

A Line. Refers to cytoplasmic induced male sterile line which is being used as female parent in hybrid seed production. *See also* **B. Line.**

Abiogenesis. An old idea, now completely ruled out that living things may be produced from non-living objects *i.e.*, spontaneous generation.

Abiology. Means science of non-living things.

Ablactation. The cessation of milk production by mammary glands when an offspring is weaned.

Abnormal Milk. The term used for the milk which is not smooth and uniform but flaky, ropy, stringy, watery, slimy, straw coloured, discoloured or bloody, bad odoured or flavoured. It occurs because of disease, feeds or medicines administered, rupture of blood vessels or injuries, unclean milking or utensils or insanitation.

Abnormal Seedings. The seedings lacking the capacity for continued development into normal healthy plants. Examples include seedlings with weak or malformed parts such as stubby roots, split hypocotyl, absence of primary leaves and/or terminal bud.

Abomasum. It is the fourth stomach of Ruminants which is lying close to the Omasum or third stomach, in which digestion gets completed by the supply of digestive juice. It is also known as Reed Stomach, Rennet Stomach, and sometimes called Maw.

Abortion. Means miscarriage, slippings, or premature delivery of partially developed offspring. (Contagious Abortion)

Abortion Antigen. An antigen which finds use in making the agglutination test.

Abrasion. (1) A superficial injury caused to the skin or the mucous lining (2) The wearing away of part of the earth's surface by the action of water, wind or ice.

Abrasive. Substances which are able to facilitate the machanical transmission of certain viruses. They tend to increase the number of wounds and also the type of injury conducive to successful establishment of infection.

Abscisic Acid. A plant hormone which has growth inhibiting action and promotes abscission. It is associated with the onest and maintenance of dormancy.

Abscission. (1) The term used for the shedding of leaves, foliage branches, floral parts, fruits or bark by the layingdown of an abscission layer. (2) Also, refers to the liberation of a fungus spore by the disappearance of an adjoining layer or wall.

Absolute Drought. The term used for a period of atleast fifteen consecutive days, none of which has received as much as 0.25 milimeters of rain.

Absolute Growth Rate. Defined as the rate of increase in size of a growing plant (or a part of it) in a given time, under specified conditions.

Absolute Humidity. Refers to the mass of water vapour present in a unit volume of air. It may be expressed as grams per cubic meter.

Absolute Transpiration. The term used for the rate of water loss from a plant determined experimentally.

Absolute Water Requirement. Means the quantity of water in acre-inches per crop season which is absorbed by the crop together with the evaporation from the crop producing land. Also called 'consumptive use of water'.

Absolute Weed. One of no use to farners under any circumstances.

Absorbed Water. Water which is held mechanically in a soil mass and possessing physical properties which is not substantially different from ordinary water at the same temperature and pressure.

Absorptiometer. An instrument used for measuring the absorption of nearly monochromatic radiation in the visible range by a gas or a liquid, and so is used to find out the concentration of the absorbing constituents in the gas or liquid.

Abstraction Licence. A licence issued by a water authority for the abstraction of water from a water source for domestic or commercial use, after for irrigation purposes.

Acaricides. Substances that are able to destroy acarids (ticks and mites). They find use in the treatment of acariasis.

Acarologist. Specialist in mites.

Acarology. Deals with the study of acarina which is a large order of small to microscopic arachnids including ticks and mites.

Acclerated Erosion. Refers to the erosion of soil material which takes place at a rate greater than that of normal erosin. It is caused due to the destruction of vegetative cover or due to some activity of man.

Accession. (1) Refers to a cultivar which is registered at a genetic resources centre (2) Any entry which has been newly received in a crop improvement programme.

Accession Collection. Refers to the collection in which seed of every entry has been saved. This reduces the chance that genetic loss will take place through elimination of apparent 'duplicates' and the masking of useful traits by bulking.

Accessory Pigments (Photosynthesis). Pigments such as carotenoids and phycobili-proteins which obtain light energy and transfer it to the reaction centres.

Acclimatization. The term used for the process by which human beings, animals and plants get adapted to an unfamiliar set of climatic conditions.

Acclimatisation Value. The extra worth of a flock of sheep acclimatised to a farm, over an imported flock which may stray, with loss of sheep, (Hefted Sheep).

Acclimatised Sheep. Sheep which are long accustomed to their local environment. They will normally not stray beyond the farm boundary, unlike an imported flock.

Accredited Herds Scheme. Brucellosis (Accredited Herds) Scheme.

Accredited Milk. Milk sold by a producer-retailer from a herd accredited as free from Brucellosis.

Acetonaemia. Bovine Ketosis.

Accrescent. Enlarged and persistent. Especially of a calyx which becomes enlarged as the fruit ripens.

Accretion. (1) Refers to the process of growth in which material has been added to the outside in non-living matter (2) Also, refers to the gradual addition of new land to old by the deposition of sediment carried by the water of a system.

Accumulation or Summation Curve. Curve which is obtained by plotting settling time on the abscissa and the weight of the deposited particles on the ordinate. It is representing the total amount of accumulated material at various time intervals.

Acenapthene. Chemical agent which is able to induce polyploidy, $C_{10}H_6(CH_2)_2$.

Acentric. Whole or part of a chromatid or chromosome which lacks centromere.

Acephalous. Lacking a head. Said of a style lacking a well developed stigma.

Acetabularia. It is the genus of tropical or subtropical marine algae of the division Chlorophyta.

Acetic Acid Bacteria. Bacteria which are able to produces acetic acid from alcohol.

Acetobacter. It is the genus of gram-negative, chemoorganotrophic, aerobic, ovoid or rod shaped non-spring bacteria. They are able to oxidize ethanol to acetic acid and acetic acid to carbon dioxide.

Acetomonas. It is the genus of gram-negative, aerobic, chemoorganotrophic, ovoid or rod shaped polarly flagellate (or non-motile) non-spring bacteria belonging to the family Pseudomonadaceae.

Acetonemia. The term used for the metabolic disorder of cattle which causes an abnormal quantity of ketone bodies in the blood and urine. It is responsible for a distinctive flavour to milk. It is associated with poor quality of ration.

Achene. A small, dry, one-seeded Fruit formed from a single Carpel, which does not split open on ripening but rots gradually with seed germination. The pips on strawberries are achenes.

Achrodextrins. Dextrins which are produced when the hydrolysis of starch has carried out to the stage that it imparts no colour with iodine.

Achromatin. Non-staining basic substance of the nucleus excluding the chromatin.

Acid Fast Bacteria. Bacteria keeping carbol fuchsin stain after addition of 25% sulphuric acid.

Acid Fast Organisms. Micro-organisms which, once effectively stained, cannot get decolorized when treated with mineral acids or mixtures of ethanol and acid.

Acid Forming Fertilizer. A fertilizer leaving acid residue in the soil when applied.

Acid Scarification. Its purpose is to modify hard or impermeable seed coverings. Soaking seeds in concentrated sulphuric acid is considered to be an effective method of doing this.

Acid Soil. Soil having a preponderance of hydrogen ions, and probably of aluminium in proportion of hydroxyl ions. Specifically, it is a soil with a pH value less than 7.0.

Acidosis. The term used for the pregnancy disease of animals which involves a metabolic disorder, a decreased alkalinity of blood and tissues due to the presence of excessive amount of acids in the system.

Acinetobacter. It is the genus of gram-negative, aerobic, chemo-organotrophic, asporogenous, aflagellate bacteria of the family Neisseriaceae.

Acquired Immunity. Means specific immunity which is acquired through exposure to the antigen.

Acre. A measure of land. 4840.sq. yds. Equal to 4 Roods, or 10 square Chains. (Scottish acre, 6150.4 sq. yds; Irish acre, 7840 sq. yds- both obsolete.) Also equivalent to 0.405 Hectares. Originally, the amount of land that could be ploughed in a day by a team of oxen.

Acre Foot of Water. Defined as the amount of water that would cover an acre of land to a depth of one foot, provided that there occurs no seepage, evaporation and runoff losses.

Acre-inch. A unit of measurement used in Irrigation works. The amount of water required to cover an acre to a depth of 1 in.(25 4 mm per hectare) equivalent to 101 tons or 22460 gallons.

Actinobacillosis. An animal disease mainly of cattle and pigs. It is caused by bacteria (*Actinobacillus lignieres*) causing swelling and hardening of tongue and face. It is also known as Wooden Tongue. The disease is also found in sheep and is sometimes called 'Cruels'.

Actinomycin D. An antibiotic that is able to inhibit the elongation of RNA chains.

Actinomycosis. An animal disease which is caused by the microscopic 'ray-fungus' (*Actinomyces bovis*) that penetrates through small mouth wounds (caused by barley awns, foreign bodies, during change of teeth, etc.) resulting in the swelling of the jawbone with ultimate suppuration. Also known as Lumpy Jaw.

Activated Sludge. Refers to the thick or thin watery suspension of sewage and associated protozoans, micro-metazoans and bacteria in certain sewage treatment processes. When the suspension is constantly stirred and air is blown through it, the organisms convert much of the organic fraction into gases, salts, and inert organic compounds.

Active Immunity. Refers to the specific immunity which is afforded by the body's own immunological defence mechanism following exposure to antigen.

Active Ingredient. Refers to an active component of a formulated product like a fungicide. For example Ridomil 25 WP has 25% of the active component acylalenine and 75% inert material. The ridomil has 25% a.i.

Acute Exposure. Term used for radiation exposure of short duration.

Adaptive Enzyme. An enzyme which is able to synthesize only in the presence of an inducing agent, such as its substrate.

Additives. (1) Substances added to Feedingstuffs or Concentrates during manufacture, which are not principal nutrient sources. Examples include Vitamins, minerals, Trace Elements, certain Antibiotics, etc. (2) Substances added to pesticides to improve their efficiency and efficacy. They include Surfactants (wetters), Adjuvant Oils and drift retardants.

Additive (Fertilizer). A substance which is added to fertilizer to improve its chemical or physical condition.

Additive (Silage). Any material which is added to forage at the time of ensling.

Addled Egg. The term used for a fertile egg in which the embryo died between the seventh and fourteenth day of incubation. Also a rotten egg in general terms.

Adjuvant. A material which is added to an insecticide to aid its action. Adjuvant may act in various ways like emulsifying agents, wetting agents, spreaders and stickers.

Adjuvant Oils. Vegetable-based and mineral oils which are used to improve the effectiveness and reliability of pesticide sprays. The oils reduce evaporative loss from droplets in flight and help maintain the chemical in liquid phase for extended periods on leaves, aiding adsorption.

Adsorption Complex. The group of substances in the soil which are capable of adsorbing other materials (such as applied fertilizers). Organic and inorganic colloidal substances form the greater part of the adsorption complex. Non-colloidal materials, such as sand and silt, exhibit adsorption to a much lesser extent than the colloidal materials.

Adulteration. The adding of water to milk.

Adventitious. Arising in an abnormal position. Adventitious roots develop from parts of plants other than roots, *e.g.*, from stem or leaf cuttings. Adventitious buds develop from parts of plants other than in the axils of leaves, *e.g.*, from roots.

Adze. A tool consisting of a thin arched blade with its edge at right-angles to the handle like a hoe, used for slicing off the surface of a piece of wood, *e.g.*, in fencing work.

Aeciospore. A yellow, single-celled, binucleate spore of the rust fungus which is formed in a special cluster or cup-like structure called the accium.

Aecium. A cup like structure which is produced on certain plants by rust fungi, in which binucleate single-celled spores are found.

Aeolian Soil. Wind deposited soil which is easy to keep in good tilth, as coastal sand dune, of not much agricultural value. It occues in parts of Rajasthan, South-east Punjab and Kutch.

Aerial Photograph. Refers to the photograph of the earth's surface which is taken from airborne equipment. It is sometimes called aerial photo or air photograph.

Aerobe. Any organism glowing in the presence of oxygen or air.

Aerobic Respiration. Refers to enzymatic destruction of a substrate to release energy, by using elemental oxygen, evolving carbon dioxide and water.

Affluent. Stream which is flowing into a larger stream; a tributary.

Afforestation. The process of transforming an area into forest, usually when trees have not previously been grown there. (Deforestation, Reforestation)

Aflatoxin. A poisonous toxin which is produced by the fungus *Aspergillus flavus* and found in Ground Nut Meal. However, the fungus is also found on cotton seed, palm kernels and even in maize. The toxin causes a reduction in milk yield and growth rate in cattle, and sometimes jaundice in pigs eating the meal. It is also passed on through the animal's feed to the consuming public.

African Coast Fever. Rhodesian Tick-Fever.

African Horse-Sickness. Refers to an infectious disease of equines, which is characterized by market oedema of subcutaneous tissues and lungs, haemorrhages in some of the internal organs and accumulation of serious fluid in the body cavity. It has been caused by filterable virus having a particle size of about 50 m.

African Pig Disease. Wart Hog Disease.

African Swine Fever. Wart Hog Disease.

African Trypanosomiasis. Refers to Nagara.

After Crop. A second crop from the same land in the same year.

After- cultivation. Harrowing, rolling and other cultivations carried out in a field after the crop has emerged.

Aftermath. Grass springs up again after the cutting of a crop of Hay, and can subsequently be taken as a second cut. Also known as eadish, eddish, rowen and lattermath.

After Math. Refers to recovery growth of forage plants after harvesting, either by animal or by machine.

After Ripening. The term used for a period through which some seeds have to pass after ripening, before they will germinate.

Agalactia. It is also known as suppression of milk. It has been bound to be neither as infectious nor as common in cattle as in sheep and goat. It usually takes place at calving time. It is caused by such predisposing causes such as indigestion, loss of apetite, mastities, inadequate or poor quality feed, plant poisoning, thirst, enforced driving, fear excitement or the removal of the young.

Agamospermy. Refers to the reproduction in which meiosis and fertilization get circumvented, so that the embryo (s) which develops in the seed has been usually indentical with its maternal parent in both chromosome number and genic content. It includes all types of apomictic reproduction in which embryos and seeds have been formed by asexual means.

Agar. A complex sulphated galactan which is used as a base for several solid and semi-solid media used in microbiological studies. It is formed by certain species of red algae.

Agglomerated Feeds. 1. Refer to the compacted or extruded form of individual ingredients. 2. Also, refers to a mixture of non-processed and/or individually processed ingredients or a combination of both.

Agglutination. The term used for a process in which cells or bacteria undergo cross linking with or gets attached to each other when the antigens on the surface of their cells undergo intersection with antibodies, which form bridges linking the antigen determinant sites of the different cells. Such reactions are used for identifying blood groups, bacteria etc.

Agistment. A contract arising when one person, the agister, takes livestock belonging to another person to graze on his land for reward.

Agmark. It is a sort for agricultural marketing, national insignia for quality and purity and applicable to agriculture and animal husbandry products, under the Agricultural Produce Act, 1937. Labels carrying the mark assigned to products conform to certain grade specifications and standards laid down.

Agricultural Chemistry. Refers to that branch of the science of chemistry which deals with the composition and transformation of the plants and animals on which the economy of the farmer depends.

Agricultural Economics. Branch of economics which deals with farm management and production.

Agricultural Engineering. (1) Deals with the application of knowledge, techniques and disciplines of various fields of engineering to the solution of problems which arise in the field of agriculture and rural living with the aim of reducing labour, improving agricultural productivity per worker, raising the standard of living of farmers and increasing the overall earnings per worker (2) Deals with the study in the design, construction and use of agricultural implements and buildings, soil and water management, use of electricity and processing of agricultural products.

Agricultural Geology. A branch of applied geology which is concerned with the nature and properties of rocks and minerals of importance of soil formation.

Agricultural Marketing. Includes all the services or functions which are involved in bringing produce from a farmer to the final consumer.

Agricultural Science Service. A division of the agricultural development and advisory service providing specialist advice to farness on a range of scientific disciplines both directly and via field officers.

Agriculture. Science of cultivating soil so as to produce economic crops and demanding great knowledge and skill in its scientific, commercial and artistic aspects. Very broadly the term is applicable to include both pastorcil and arable farming, even though the trend is some what towards the latter.

Agrobiology. Phase of the study of agronomy which deals with the relation of yield to the quantity of an added or available fertilizer element.

Agroceric Acid. $C_{26}H_{44}O_3$. Refers to a constituent of humus.

Agroecology. Refers to the study of the relation of agricultural crops and environment.

Agrology. Refers to the study of applied phase of soil science and soil management.

Agronomy. The branch of agriculture concerned with the theory and practice of field-crop production and the scientific management of soil.

Agrosterol. $C_{26}H_{44}O$. It is a constituent of humus.

Agrostology. Deals with the study of grasses, their classification, management and utilization.

Air Blast Sprayer. Sprayer having a large fan to produce a high speed, high volume air flow to break spray particles into small droplets and to carry the spray to the plant.

Air Drainage. Refers to the flow of cold air downhill. Freeze-sensitive crops are grown on hillsides so that on calm spring nights the cold air will drain down and away from the crop.

Air Dry Soil. Soil losing its free moisture by evaporation or transpiration, and therefore having only hygroscopic moisture and chemically combined water.

Air Layering. It is a method (asexual) of plant propagation by layering wherein roots are formed on the aerial part of a plant after the stem gets injured or girdled or slit at an angle. This portion is then kept in a moist rooting medium. It is commonly used for propagating a number of tropical and subtropical trees and shrubs.

Air Screen Cleaner. Basic seed processing machine in which basic cleaning of all seeds is done by air screen cleaner.

Air Space Ratio. Refers to the ratio of volume of water which can be drained from a saturated toil under the action of force of gravity to total volume of voids.

Air Void Ratio. Refers to the ratio of the volume of air space to the total volume of voids in a soil mass.

Aitionomic. The term used for the ability to develop parthenocarpic fruits only in response to some stimulus external to the ovary.

Albedo. Refers to the ratio of the amount of radiation reflected to the amount coming on a surface. The earth's mean albedo has been rearly 35 per cent.

Albinism. Means the condition in which the normal pigment is not present, such as in the skin, hair and eyes. In plants also lack of normal chlorophyll causes albinism.

Albino. An animal with no skin pigment, white-coated, pink-eyed. Commonly due to a single recessive gene. (Dominant Gene)

Albumens. A group of water-soluble Proteins which occur in many of the tissues and body fluids of animals ; *e.g.*, egg albumen (egg white); serum albumen (in blood), lactalbumen (in milk), etc. Also called albumins.

Albumins. See albumens.

Albuminoids. The simple proteins which are insoluble in all neutral solvents and in dilute acids and alkalies but are coagulated by heat. For example keratin, collagen etc.

Albuminous Seed. Seed having endosperm at the time of germination, which nourishes the growing seedling.

Aldrin. A persistent organochlorine insecticide, a derivative of chlorinated nupthalene, harmful to fish and under restricted use. Applied as a dust, spray or seed dressing.

Aleurone Grains. Protein bodies which are present in certain oilseeds. They consist of ground substance having one or more of the crystalloids or protein crystals, calcium oxalate crystals, globoids believed to be phosphates with organic salts.

Aleurone Layer. Layer of aleurone grains which is presents in the periphery of most seeds.

Aleuroplast. A plastid storing protein, often in the form of grains.

Alfalfa Gate. In irrigation, it refers to a light sheet metal slide gate in a section of pipe for the control of water from sublaterals into fields and ditches.

Alfalfa Meal. A feedingstuff comprising artificially dried and ground Lucerne, known in America as alfalfa.

Algae. Simple, flowerless, unicellular or multicellular photosynthetic plants with unicellular organs of reproduction. They are aquatic plants, freshwater or marine, (*e.g.* seaweeds), or plants of damp situations, (*e.g.* damp walls, tree trunks, in soil, etc.)

Algicides. Chemical agents which are used for killing the algae (selectively).

Alkali Flat. The term used for on alkaline, marshy area in an arid region in which one or more direct streams lead. It becomes a barren area of hard mud which gets covered with alkali when all the water is evaporated in the dry season, after heavy rainfall, it becomes a shallow muddy lake.

Alkaline Soil. Sweet Soil. Soil having a pH above 7.0 due to an excess of hydroxyl (OH^-) ions, or having a high exchangeable sodium content, or both. Typical alkaline soils are those of the chalk and limestone hills, e.g., South Downs, Cotswolds and Craven Pennines. (Acid Soil, Neutral Soil).

Alkaloids. Heterogenous group of basic nitrogen having substances that are produced by plants and possess potent pharmacological activities. They have been classified according to the type of heterocyclic group present.

Allantois. A membranous sac-like appendage for effecting oxygenation in the embryos of mammals, birds and reptiles.

Alleyways. Passage-ways between rows of hop plants in hop gardens.

Allochronic Speciation. Refers to the production of new species during the passage to time, tending to form a gradation from one species to the next.

Allogamy. Mode of reproduction in which a flower gets pollinated by pollen from another flower.

Allopatric. Refers to two or more related populations which are occupying mutually exclusive (but usually adjacent) geographical areas and do not interbreed; this situation is often termed as contiguous allopatry; where two related populations get separated by a wide gap in which neither occurs, the situation is known as disjunct.

Allopatric Hybridisation. Refers to the hybridisation between incompletely differentiated species in a border zone, because of a premature break-down of a geographical barrier.

Allopolyploid. A polyploid organism to which two different species have each contributed one or more sets of chromosomes (e.g., probably cultivated wheats).

Allotetraploid. An allopolyploid arising when an ordinary Hybrid between two species which contains a chromosome set from each parent, doubles its chromosome number. Ordinary hybrids are usually sterile since their chromosomes cannot pair during meiosis. In an allotetraploid each chromosome can pair with its homologue, thus overcoming sterility, and immediately creating a new species. Tobacco probably originated in this way. Allotetraploidy is only known in plants at present. Artificial production of allotetraploids is important in creating new agriculture and horticultural varieties. Also called amphidiploid.

Alluvial. The term used for the material that gets transported and deposited by the running water.

Alluvial Deposit. Soil, silt, sand, gravel, or other material which has been deposited in place by the action of running or receding water.

Alluvial Fan. The outspread sloping deposit of detril material which has been brought down by the action of water from neighbouring elevations upon a plain or open valley bottom and dropped where the stream gradient decreases abruptly, forming roughly a segment of a low cone with the apex at the point of debouchure.

Alluvial Land. Areas of unconsolidated alluvium which are generally stratified and varying widely in texture, recently deposited by streams, and subject to frequent flooding. A miscellaneous land type.

Alluvial Soil. Soil which is formed by the transportation and deposition, by streams, of material which is carried to a considerable distance before getting deposited.

Alluvium. Material such as clay, silt and sand deposited by rivers and streams. (Colluvium).

Alpha Amylase. An enzyme naturally occurring in wheat seed at germination, causing starch breakdown to sugar. If unduly active in milling wheat it reduces the quality of bread and cakes. Tested for by either the Hagberg Test or the Farrand Test.

Alpine. A breed of goat native to the Swiss Alps, from which the British Alpine breed developed.

Alsike Clover. A species of clover, *Trifolium hybridum*, named after Alsike near Uppsala in Sweden, characterised by pale pink or white flowers, a smooth fairly upright stem, toothed leaflets with long pointed stipules, and a pod containing 1-3 small seeds. It tolerates adverse soil conditions better than Red Clover or White Clover, and is less susceptible to stem rot.

Alternate Bearing. The term used for the habit of bearing heavy and sparse crops in alternate years, inherent in mango and some other species of fruit trees, where terminals bearing fruit in an year put forth only leafy shoots next year.

Aluminium Phosphide. A chemical which acts as a funmigant for grain. Available commercially as tablets; 4-5 days exposure allowed.

Amel Corn. An inferior variety of wheat (*Triticum vulgare dicoccum*), the larger spelt. Also called French Rice.

Amelioration. The process of enhancing the agricultural value of land by drainage, tillage, liming, manuring, etc.

Amendment. Any material like lime, gypsum, saw dust, sulphur or soil conditioner which is added to the soil to influence plant growth by improving the physical and biological properties of the soil.

Ammonia (NH_3). A pungent colourless gas used in the manufacture of fertilizers (e.g., Ammonium Nitrate, Sulphate of Ammonia).

Anhydrous ammonia is a very concentrated nitrogenous fertilizer (82% nitrogen) consisting of pure ammonia in liquid form under pressure and applied by injection into the soil, usually when moist, where it vaporises and combines with soil eolloids. Aqueous ammonia, a by-product from gas works called gas liquor, comprising ammonia in solution, used to be commonly used as a fertilizer, applied to the soil surface.

Ammonium Fixation. The adsorption or absorption of ammonium ions by the mineral or organic fractions of the soil so that the ions become relatively insoluble in water and unexchangeable by cation exchange.

Ammoniated Mercury. A white poisonous powder which is soluble in acids. It finds use as antiseptic and parasiticide, especially in the form of ointment having about five per cent ammoniated mercury.

Ammoniated Superphosphate. A fertilizer having 5 parts of ammonia to 100 parts of superphosphate.

Ammonification. Means the decomposition of nitrogenous organic matter by micro-organisms, with the release of ammonia.

Ammonium Nitrate. (NH_4NO_3). The commonest nitrogen fertilizer currently in use either as a straight fertilizer or in compound fertilizers. Now available as a pure salt, safe for agricultural use, but prior to 1965 it was only available as an unstable synthetic salt, liable to caking, and was usually blended with chalk or ground limestone. (see nitro-chalk).

Ammonium Phosphates. Mono and di-ammonium phosphates are occasionally used in the pure state as fertilizers but, more commonly, they are constituents of concentrated compound fertilizers. Both the phosphate and the nitrogen are soluble in water (see fertilizers, artificial fertilizers).

Ammonium Phosphate Sulphate. A product obtained by treating a mixture of phosphoric acid and sulphuric acid with ammonia. Consists principally of a mixture of ammonium phosphate and ammonium sulphate. Used as fertilizer.

Ammonium Sulphate. A white crystalline salt having about 20 to 21 per cent nitrogen, widely used as fertilizer. The nitrogen is not easily lost in drainage, because the ammonium ions replace calcium in the exchange complex and thus gets retained by soil particles. It is suitable for wetland crops like paddy and jute, also for other crops grown on variety of soils. It has acidic effect on the soil.

Amnion. A protective membranous sac containing amniotic fluid cushioning the developing embryo of an animal.

Amniota. Any reptile, bird or mammal, their embryos possess an amnion and an allantois.

Amphidiploid. An allopolyploid which is obtained by hybridization between two plant species or genera where the chromosome set contributed by each parent undergoes doubling and produces in the hybrid a chromosome number which is the sum of diploid numbers of the two parent forms.

Amylase. Enzyme acting on starch.

Amylopectin. A polysaccharide which is made up of branched chains having a large number of glucose units linked in the 1-4, 1-6 position. A constituent of most starches.

Amyloplast. Plastid capable of storing starch, usually as grains. It occurs in the endosperm of storage organs, tubers, roots and seeds

Amylopsin. An enzyme found in pancreatic juice which breaks down starch into sugars.

Anaerobes. Organisms capable of living in absence of free oxygen (gaseous or dissolved).

Anaerobic Conditions. Environmental conditions in which free oxygen (gaseous or dissolved) is absent.

Anaplasmosis. An infectious disease of cattle, characterised by anaemia.

Analgesia. The term used for the absence of sensitivity to pain. This can be produced by medication given for the relief to pain and can also happen in some physical and emotional disorders.

Analgestic. Drug which lessens or prevents pain. For example morphia.

Analytic Seed Sample. Seed sample which is obtained by dividing the mechanical seed sample in which the analysis for different tests is done conveniently.

Anaphylaxis. Refers to a state of increased sensitivity or hypersensitivity in an animal which follows the injection of an antigen as a protein.

Anaplasma Marginal. Minute parasite which is responsize for anaplasmosis in the animals.

Anaplasmosis. Refers to a serious, infectious blood disease of cattle which is not transmissible to man, but goat, sheep, buffalo, camel etc, are susceptible. It is caused by a minute parasite *Anaplasma marginale* which invades and destroys large number of red blood cells. The infection occurs through ticks, mosquitoes and other biting insects. Symptoms include rise in temperature from 103°F to 107°F and pulse rate become to 70 to 140 a minute. The red blood cell count decreases from a normal of about 5 to 7 millions per cent of blood to 2 millions or less in severe cases. The sick cattle are generally constipated and abortion often takes peace in advanced pregnancy. For treatment, sick animals should be kept in the shade and given plenty of water and should be protected against flies and the biting insects. In early cases potassium arsenate solution have been found to be useful.

Anatomy. Science of morphology (structure of organism) and usually but not always, refers to the parts seen without a microscope.

Anemia. A condition in which the blood becomes deficient either in quantity or in quality. The deficiency in quality may involve diminution of the amount of haemoglobin or in dimnution of the number of red blood corpuscles.

Anemophilous Plants. Plants which are pollinated by the wind.

Anemophilous Plants. Plants which are pollinated by the wind.

Aneurism. Means a spindle shaped, globular or cylindrical dilation of a blood vessel. It often has a heavy deposit of fibrin causing an interference with the circulation of the blood. Immature forms of one of the palisade worms is known to cause aneurism in horses.

Angiosperms. The higher flowering plants in which the flowers are more complicated and the seeds get enclosed in the fruits. They are of two types, (a) Dicotyledous-embryo of seed bearing two cotyledens (b) monocotyledous-embryo of seed bears only one cotyledon.

Angiospermy. Refers to the condition in which seeds are enclosed within an ovary—a characteristic of flowering plants(angiosperms) that distinguishes them from all other plants.

Anhydrase. An enzyme which is capable of accelerating the removal of water from a substance *e.g.*, carbonic anhydrase removes water from carbonic acid.

Animal Fats. Fats derived from animals and characterised by a high proportion of saturated fatty acids, as distinct from plant fats or oils which are characterised by a high proportion of unsaturated acids.

Animal Husbandry. Deals with the practice of scientific methods of animal breeding, feeding management and disease control for getting the best economic returns from them.

Animalize. The chemical treatment of vegetable fibres, such as cotton or flax, so that the fibres will resemble animal fibres, like wool, in their behaviour toward wool dyes.

Annual. A plant which completes its life cycle in one year, *i.e.*, grows from seed, flowers, fruits, and dies.

Annual Ring. The annual growth of secondary wood (xylem) in stems and roots of temperate woody plants, seen as a series of concentric rings in a stem cross-section due to the distinction in size between xylem elements formed in spring and autumn. Annual rings are used to estimate the approximate age of a tree. Growth is constant throughout the year in the tropics and no rings are seen. Annual rings may also be seen in cow horns.

Antacid or Antiacid. Medicine which is able to counteract acidity *e.g.*, sodium bicarbonate. Antiacids are used to relieve indigestion and other conditions due to hypersensitivity. Some antiacids possess laxative effect, when used too often.

Antagonistic Medicine. Medicine which is able to counteract the action of another medicine.

Antecedent Moisture. Refers to the degree of wetness of the soil at the beginning of a run off period frequently expressed as an index determined by summation of weighed daily rain-falls for a 10 to 20 day period preceding the runoff in question.

Antecedent Soil Water. Refers to the degree of wetness of the soil prior to irrigation or at the beginning of a runoff period. It is usually expressed as an index or as total inches of soil water.

Anthocyans. Water soluble pigments of higher plant vacuoles, having the anthocyanidin flavonoid pigments and their glycosides, the anthocyanins. They are mainly responsible for the red, blue and purple colours of flowers, fruits, leaves, and buds.

Anthrax. A notifiable disease of animals caused by a bacterium, *Bacillus anthracis,* which lingers in the soil and in infected carcases. It occurs world wide but is endemic in Europe. In Britain it occurs in cattle and pigs, rarely in sheep and horses, and the chief source of infection is imported infected feed-usually from India and African countries. The disease is characterised by high fever, sometimes diarrhoea, and death in 24-48 hours with blood usually emanating from the body orifices. There is a legal requirement for carcases to be burnt.

Anthrax Bacterin. It tends to stimulate the treated animal to produce immune bodies *i.e.,*active immunity. Its anthrax bacterin being sterile, is in itself incapable of producing disease in the treated animal, it is, therefore, safer than any of the anthraxspore vaccines.

Anthrax-Spore Vaccinations. These consist of living anthrax spores. They are used alone or in combination with anthrax serum. The spores in these vaccines become so weakened that they will not produce the disease in livestock under ordinary conditions, but some susceptible animals may die. For this reason these vaccines should be used only on those premises where the disease has occurred earlier.

Anthrax Vaccination. The term used for a preventive vaccine against a serious disease of livestock called anthrax.

Antianthrax Serum. It is responsible for increased resistance to anthrax in direct proportion to the quantity of serum which is given to the animal. It is of value both as a preventive and as a therapeutic agent. It should be used as a preventive only when immediate protection is needed because the immunity which it confers is of relatively short duration.

Antibacterial Agent. Agents which are able to kill or inhibit the growth of bacteria *e.g.*, antiseptics, antibiotics and disinfectants.

Antibiotics. Antibacterial drugs obtained from living organisms, such as moulds, bacteria or other micro-organisms, which act by inhibiting the growth and multiplication of germs and other bacteria in the body. Examples are penicillin and the tetracylines. Antibiotics have no effect on illnesses caused by viruses.

Antibiotic Feed Supplement. A feeding material which is used chiefly for its antibiotic content *e.g.*, terramycin, auremycin, penicillin and steptomycin. Thus it is a single or a combination of antibiotics having growth-promoting properties. They give rise to a more rapid growth by altering the flora in the digestive tract of the animals.

Antiblackleg Serum. A serum which is used for treating calves which are already affected with blackleg as well as for preventing it. It is produced by injecting a pure culture of blackleg organisms into the veins and later under the skin of horses and drawing the blood afterwards.

Antibody. A natural substance produced by the body in response to the presence of foreign substances (antigens), such as bacteria or viruses, to protect itself against disease or infection. The antibody neutralises the antigen by combining with it. Most vaccines work by stimulating the formation of antibodies against a particular disease.

Antidotic. A chemical substance which is produced by microorganism and has the capacity to inactivate the injurious effect of bacterial toxins.

Antigen. A proteinous substance which is capable of stimulating the production of neutralizing substance, when it is injected into a vertebrate.

Antigen-Antibody Reaction. Any reaction which takes place between an antigen and its corresponding antibody it may involve agglutination, precipitation, complement-fixation etc., the sensitivity of allergic reactions of many individuals towards certain kinds of foods, pollen, dust, feathers, hair, drugs, etc., are all examples of antigen-antibody reactions.

Antiinfective. A remedy which is administered to overcome an infection in livestock *e.g.*, sulfona-mides, penicillin, etc.

Antimycin. An antibiotic functioning as a respiratory inhibitor in mitochondria, but generally not in bacteria.

Antipruritic. An agent which is used for relieving itching.

Antipyretic. A remedy which is able to lower the body temperature.

Antirheumatic. Drug which is used to prevent or cure rhemuatism in animals *e.g.*, sodium salicylate.

Antiseptic Dusting Powder. Powder having boric acid, sodium perborate or iodoform induces healing of wounds and also prevents attacks of flies.

Antiseptics. These are chemical agents which are used to treat human and animal tissues, usually the skin, with the object of inactivating or killing those microorganisms which are causing an infection, *e.g.*, acriflavine bisphenols, ammonium compounds. etc.

Antiserum. Blood serum of an animal which has antibodies.

Antispasmodic. A remedy which is used for relieving spasm.

Antitoxin. A type of Antibody that acts against a poison or toxin that has entered the body.

Alphid. A plant house, also called greenfly. Aphids are small, soft-skinned, often green plant-bugs. They suck sap mainly from young leaves and shoots, causing reduced growth and vigour, and leave a sweet, sticky excretion (honeydew) which is eagerly eaten by ants and which occludes the leaf and shoot pores. Often found in large numbers on wild and cultivated plants. Many are pests, *e.g.*, Bean Aphis.

Apiary. A place where Beehives ar kept.

Apiculture. Bee-keeping.

Apomixis. Refers to the development of an individual from an unfertilized egg without involving sexual fusion, whether the egg be normally haploid or abnormally diploid through failure of reduction division. Broadly may include also asexual propagation by various methods.

Aposematic Coloration (Warning Coloration). The brilliant and striking patterns of colours which occur in insects especially those which ae predator resistant, which results in a predator learning to avoid the species.

Appetitive Behaviour. Any pattern of animal behaviour that results in the satisfaction of a specific need, *e.g.*, searching by an animal for food or a mate.

Apple Blossom Weevil. An insect (*Anthonomus pomorum*), approximately 1/2 cm long, laying its eggs in apple and pear flower buds, on which the larvae feed so that no fruit develops.

Apple Sawfly. Sawfly.

Apple Sucker. A plant louse, *Psylla mali*, able to jump with its large hind legs. Its pale yellow, stump-winged offspring suck sap from opening apple flowers, causing fading and browning.

Applied Insect Control. Refers to the use of sprays, insecticides and other methods in insect control.

Approach Grafting. Refers to a type of grafting wherein two independent, self-sustaining plants have been grafted together. The top of the stock plant has been removed above the graft and the base of the scion plant has been removed below the graft, after a union has taken place. It is a method of establishing a successful union between certain plants which are difficult to graft together otherwise. Approach grafting can be carried out at anytime of the year. The cut surface should be securely fastened together and covered with grafting wax.

Aquifer. A bed of rock which holds water and allows water to percolate through it. (Artesian Well).

Arable. Land which is suitable for the production of crops that need fillage.

Arable Crops. Annual crops which require the land to be ploughed and cultivated (unless direct drilled) and seeds to be sown, as distinct from those which yield a product without such annual activities (*i.e.*, fruit, grass, etc.). The main arable crops grown in the India include the cereals, potatoes, rootcrops, and various vegetables.

Arable Land. Land capable of being ploughed, that is, fit for tillage, as opposed to that only suited for pasture. The term generally means land ploughed at regular intervals.

Arachnida. A class of Arthropods including spiders, ticks, mites, harvestmen and scorpions. They are mainly carnivorous, usually with six pairs of limbs, four for walking and the first pair as jaws. They have no antennae, simple eyes and the head and thorax are fused together.

Arboriculture. The systematic culture and care of trees and shrubs.

Are. A unit of metric land measure, 100 sq. metres. (Hectare)

Arenaceous. Sandy, Arenaceous soil have a high proportion of sand particles. Sandstone is an arenaceous rock. (See Argillaceous)

Argente. A breed of rabbit of French origin, with a silvery coat. There are several shade varieties, *e.g.*, Argente Blue, Argente Silver, etc.

Aegillaceous. Clayey. Argillaceous soils have a high proportion of clay particles. Mudstones and marls are examples of argillaceous rocks. (See Arnaceous)

Arid. A term applied to land that is dry, parched and often devoid of vegetation. Also means deficient in rainfall - usually applied to a climate or region where rainfall is barely sufficient to support vegetation.

Ark. 1. A corn bin. 2. A poultry house, usually mobile, and used for the out of doors rearing of birds. Triangular in cross-section, sometimes with a slatted floor, usually with one end covered for roosting purposes. Less common than they used to be. Also a movable shelter for pigs kept under Free-Range. (See Piggery)

Armsby Feeding Standard. In this feed standard the feeding value has been expressed in net energy which is found out in heat unit. The net energy may be defined as the metabolizing energy minus the energy lost in the heat increment or so called work of digestion. It may also be known as the gross energy minus the energy lost in faeces, urine, combustible gases and in the work of digestion. The metabolizing energy of the principal food constituents has been found out from the following.

The metabolizing gross energy per kg of each	% of proteins × 2133 calories. % of carbohydrate × 1707 calories. % of fat × 4000 calories.

The energy actually used for metabolic purposes is termed as *available energy or metabolizable energy*.

ME or AE=[Gross energy]-[chemical energy rejected]

The chemical energy rejected has been the energy of the faeces, urine and methane etc.

Arrish. Stubble or a stubble field.

Arsenic. A chemical element, steel-grey, brittle and crystalline. Found naturally in a number of minerals. Compounds are very poisonous and are used as weed-killers and pesticides. (Lead Arsenate)

Arterial Drain. A watercourse which carries water draining from smaller ditches.

Arterial Drainage. Drainage work on major rivers and water-courses which are under the control of Regional Water Authorities, Internal Drainage Boards or local Authorities. (Field Drainage)

Artesian Well. A well sunk into an aquifer sandwiched between impermeable strata in which the hydrostatic pressure normally forces a continuous flow of water up the well. The pressure is due to the well outlet being below the level of the water source. Normally occurs in basins (*e.g.*, the London Basin).

Artesian Well Capacity. Refers to the take at which a well will yield water at the surface because of artesian pressure.

Arthritis. Refers to the inflammation of a joint. It results in an enlargement and impaired movement of the joint, causing extreme lameness, often accompanied by fever.

Arthropod. The Arthropoda from the most numerous phylum of the animal kingdom. Characterised by an external skeleton of hardened cuticle, divided into segments some of which bear jointed paired limbs (the word arthropod means jointed feet). Includes insects, spiders, millipedes, centipedes, crabs and shrimps, etc.

Articulation. A joint in a stem or fruit, at which natural separation takes place.

Artificial Brooding. Means handling of baby chicks without the aid of hen. It is carried out by using heated brooder. It has advantages over natural brooding. Brooding chicks are possible at any time of the year and in large numbers. It makes control of sanitary conditions, regulation of temperature and planned feeding possible.

Artificial Hatching. Means the incubation of eggs by methods other than by natural hatching. It is usually carried out by using machines called incubators, more advantageous than natural hatching.

Artificial Insemination. The introduction, by an instrument, of semen collected from a male into the uterus of a female for animal breeding purposes. Called A.I. for short. A bull produces 50-100 times more semen in a normal mating than is needed for a cow to conceive. Thus, by the collection, dilution, and refrigerated storage of a bull's semen, it is possible for that bull to sire, via A.I., far more progeny during and even after its lifetime than would be possible by normal mating. A.I. also reduces the spread of venereal disease and enables farmers to economise by dispensing with their own or a commercial bull. Diluted semen is stored at -196°C., after the addition of glycerol, and may be kept for years.

A.I. plays an important part in cattle breeding, most of the dairy cattle in India. A.I. is now also used for sheep. Most adult rams can produce from 6000-8000 doses of fresh chilled semen, or 2500-5000 doses of deep frozen semen a year.

Artificial Parthenogenesis. Laboratory phenomena which involves the maturation of an ovum and its development into an individual without involving fertilization; the essential stimuli are shaking, pricking, heating, and treating with dilute organic acids; the phenomenon can be initiated in frogs and many marine invertebrates; in a very few instances animal produced by artificial parthenogenesis have been reared to adults.

Artificial Regeneration. Renewal of forest crop by sowing, planting or other artificial means, also the crop so obtained.

Artificial Subsurface Irrigation. The term used for the application of irrigation water below surface of soil having perforated or porous pipes which are laid underground below the root zone and water led into the pipes by suitable means. Method may make the soil saline or alkaline and the neighbouring land gets damaged due to seepage.

Artificial Vagina. Arrangement for collecting semen particularly from bull. It consists of an outer cylinder of heavy rubber casing, glass, ebonite, or metal with a thin rubber inner-tube which gets attached to the outer cylinder in such a way that a watertight space gets created between the inner and outer walls. One end of the inner tube is kept open; the tube at the other end has been tapered to fit tightly over a graduated test tube in which the semen has been collected at the time of ejaculation. There is a small valve in the outer wall for introducing warm water into the space between the inner and outer surfaces. A temperature of 41°C has been found to be near optimum for most animal species. The inner tube is well lubricated.

Artificial Vegetative Reproduction. Refers to Vegetative Reproduction.

Ascariasis. Injuries caused by the large intestinal roundworms or ascarids (large roundworms) which are present in the lumen of the small intestine. The hosts (horses, cattle, swine) obtain their parasites through swallowing their microscopic infective eggs with forage, dry feed, soil or water. Adult worms may damage the liver and lungs. The adult worms in the intestine take the lood of the host. Occasionally, they even make perforations in the intestines and produce peritonitis. Ascariasis slows down the growth of young host animals, producing more or less permanent stunting and some time even death if the worms invade vital organs. The treatment involves the administration of chenopodium oil. Swine may also be treated with sodium flouride and horses with carbon disulfide.

Ascarib. Refers to large intestinal roundworm.

Ascarops Strongylina. One of thick stomach-worm species which is occurring in swine.

Ascomycetes. A class of fungi, known as Sac Fungi, which carry spores in a sac or ascus. They range in size from very small to quite large, the fruiting bodies of the latter often being rounded or tuberous, not toadstool-shaped. Many are parasites; many cause plant diseases; others 'ripen' cheeses.

Ascon. A type of sponge having a simple canal system in which openings in the body wall led directly to spongocoel, as in *Leucosolenia*.

Ascorbic Acid. Vitamin C. An essential growth factor that is found in fruits (especially citrus) and vegetables.

Ascospore. (1) Spore which is produced because of sexual fusion subsequent to reduction division in Ascomycetes (2) Spore which is formed in an ascus and characteristic of Ascomycetes.

Ascus (Plural, Asci). (1) In Ascomycetes, it refers to a large saclike, cell, usually the swollen tip on a hyphal branch in the ascocarp within which ascospores (typically eight in number) are developed (2) Refers to a sac-like cell in the perfect stage of an ascomycete, in which the ascospores are formed.

Aseptic. Means free of microogranism which may cause contamination or infection.

Asexual Reproduction. Reproduction without Gametes. In plants it consists of either spore-formation or vegetative reproduction.

Ash. 1. Constituent of milk comprising various salt in solution, of which calcium and potassium phosphates and sodium chloride predominate, and also various vitamins (A, B, C, D and E). In normal milk the ash proportion remains fairly constant at about 0.75% (Milk Composition). A significant reduction may indicate added water. 2. A term used in the analysis of feeding stuffs for the residue left after through burning and the destruction of organic matter. Otherwise called mineral matter.

Aspartic Acid. Refers to an amino acid which is having a chemical formula $HOOC.CH_2CH_2\,CH(NH_2)COOH$. It serves as a precursor of pyrimidines, which is needed for the synthesis of nucleotide, and is an intermediate in the urea cycle.

Aspergillaceae. Family of the Aspergillales *e.g.*, Aspergillus, Penicilium.

Aspergillales. An order of the Euascomycetes which are having a close ascocarp with the asci irregularly distributed in it.

Aspergillosis. A fungal disease of chicks resulting from inhalation of spores of *Aspergillus* from damp or mouldy litter or feeding-stuffs, causing the growth of white nodules in the lungs, and resulting in breathing difficulties.

Asphyxia. Suffocation or the suspension of animation because of suffocation.

Aspirator. It is seed processing machine in which seed separations are carried out by using air on the basis of differential terminal velocity of seed. A negative pressure or vacuum is produced within the machine by the fan aligned at the discharge point. Air rushing to fill in the vaccum produces a stream of air which is used to separate seeds.

Asporogenous Yeast. Yeast which fails to produce ascospores.

Ass. A small, usually grey, long-eared animal of the horse genus. Also known as donkey. Used as a draught animal and crossed with the horse to produce a mule.

Assart. To reclaim for agriculture by grubbing. Assarted land is that cleared of trees and bushes.

Assay. Means qualitative or quantitative determination of the components of a material, as an ore or a drug.

Assimilate. Substance which is formed in plant during the manufacture of food.

Assimilation. Refers to the absorption and utilization of simple food-substances.

Assimilatory Quotient. Refers to the ratio of carbon-dioxide intake to oxygen output (CO_2/O_2). This ratio is nearly near 1, agreeing with the equation.

$$6CO_2 + 6H_2O = C_6H_{12}O_6 + 6O_2$$

Assimilatory Roots. These are long, slender hanging roots which are green in colour. These roots are capable of carbon assimilation and manufacture carbohydrates *e.g.*, submerged roots of water chestnut.

Association. A major plant community in which more than one plant species is dominant, as in mixed deciduous woodland. Also applied in modern usage to any small unit of natural vegetation (Soil Association).

Assortive Mating. Refers to the non-random sexual reproduction in a population ; tendency of certain types of males to breed with certain types of females, resulting in parents that are mentally and physically more similar than would be expected by chance.

Asthenoblosis. Refers to the theoretical explanation of the phenomenon of diapause or arrested growth of an insect. It is assumed that there occurs an accumulation of waste products in the tissues and that these restrict normal development in much the same way as they may bring about muscular fatigue. During a resting stage these substances gradually disappear because excretion is continued while normal metabolism becomes slows.

Astringent. Drug which is able to contract tissues. It is locally applied to control bleeding relieve inflammation, etc.

Astingent Lotions. Liquids having one or more astringent drugs. They often find use in the treatment of lameness in animals.

Asynapsis. Refers to the failure of chromosomes to pair at meiosis or the absence of chiasmata formation.

Atavism. The appearance in an animal of ancestral, but not parental, characteristics, usually after an interval of several or many generations. Also called reversion and loosely known as 'throw-back.'

At-foot. Suckling. Thus a ewe with lambs at foot has suckling lambs.

Athermanous. Substances which are able to absorb the radiant heat falling on them and, therefore, disallow it to pass through them are known as athermanous substances *e.g.*, water, wood, liquids, etc.

Atmometer. An instrument which is used for measuring evaporation; also called an atmidometer or evaporimeter.

Atony. Weakness of the system cr of any organ which is characterized by a lack of tone or lack of tension.

Atresia. In animals, it is a defect which is existing from birth, but it is seldom discovered until after the heifer gets freshned. Treatment is only surgical. To prevent the closure of teat orifice by healing a teat tube is inserted at milking time, and to replace it between milking with a sterile or dedicated teat dilator.

Atrophic Rhinitis. Nasal inflammation in young piglets with rapid uneven growth of the snout, causing deformity.

Attenuated Strains. Strains of pathogenic micro-organisms, mainly bacteria and viruses, which have lost their virulence.

Attenuation Weakening. Refers to the reduction in virulence. Any procedure in which the pathogenicity of a given organism gets reduced or abolished.

Attenuator Region. The term use for the region of DNA within an operon at which most RNA polymerase molecules stop transcription. Receipt of a specific antitermination factor will make transcription to take place.

Attested Area. An area in which a programme of specific animal disease eradication is operating , declared as attested, the disease having been reduced to a very low level. Specifically applied to Brucellosis and Bovine Tuberculosis.

Attractants. Refers to the materials that are able to attract insects or other animals and make them to eat or contact poison bait or sprays, and consequently cause their death.

Auger. 1. A tool which is used for boring holes. A soil auger is used to bore into the soil and withdraw a small sample for observation. 2. A mechanism which is involved for conveying granular materials (mainly grain) or slurry, comprising a power-driven helical screw in a closefitting tube.

Aujeszky's Disease. A virus disease of cattle and pigs made notifiable in 1979. It is characterised by intense itching of the hindquarters Named after the Hungarian, Aujeszky, who first described it in 1902. Cattle lick the infected part bare and may bite it. Paralysis takes place within 24 hours and death within 48 hours. In pigs the disease is milder with less mortality, characterised by loss of appetite, vomiting, diarrhoea convulsions, acute salivation, and throat paralysis. Sows may abort. The disease also affects dogs and rats which can transmit the disease by biting. Also called Pseudo-rabies. Infecticus Bulbar Paralysis, and mad itch.

Aureomycin. An antibiotic which is obtained from *Streptomycin aureofaciens*. It has been passed to be effective against both gram-negative and gram-positive micro-organisms, and certain viruses. It is administered intravenously and orally. It is also used in antibiotic feed supplement.

Autecology. Refers to the ecology of an individual organism or texonomic group, as opposed to synecology which is the ecology of a community.

Autoclave. An apparatus which uses stream under pressure for sterilization. The materials get sterilized by air-free saturated steam at temperature in excess of 100^{0}C.

Autoecism. Refers to the capability of a parasitic fungus so as to complete its entire life cycle on the same host.

Autogenous Vaccine. Vaccine prepared with germs which obtained from the patient's own blood and employed in the treatment of his disease.

Autoinfection. Means reinfection from bost's own parasites.

Autointoxication. Means self-poisoning which occurs due to the absorption of uneliminated toxins produced within the body, *e.g.*, from waste products.

Autolysis. Means breakdown of cells by the action of their own enzymes (endogenous enzymes).

Autolytic . Self-digestive. It refers to the process of self-digestion, which takes place in vegetable and animal tissues, particularly after they have ceased to be a normal part of the organism to which they belong; thus the tenderizing of beef during hanging is an example of autolytic process.

Automatic-Planter. It is a simple bullockdrawn automatic seed-cum-fertilizer planter which is useful for metering correct quantity of seed and for placing fertilizer at proper depth. It is possible to attach four such planters to tool bar of tractor. It is also equipped with a pair of maker arms for aligning the rows.

Autonarcosis. Said of the state of being poisoned, rendered dorment, or arrested in growth, because of self-produced carbon dioxide.

Autophytes. Green plants which are able to prepare their own organic food from the raw inorganic materials absorbed from soil and the air.

Autopolyploid. A polyploid organism in which all the chromosomes come from the same species. (Allopolyploid)

Autopsy. Post-mortem examination.

Auto Recorder. A milking machine comprising a container on a spring balance which records the weight of milk from each cow.

Auto-sex-linked. Pure breeds of poultry in which day-old male and female chicks respectively exhibit different colours.

Autotroph. Any microorganism using only inorganic materials as a source of energy. Environment carbon dioxide forms the main or rather the sole source of carbon.

Autotrophic. Organisms able to manufacture their own food from inorganic materials using energy from an outside source. Most green plants are completely autotrophic, manufacturing organic foods from carbon dioxide, water, and mineral salts, using sunlight energy fixed by Chlorophyll. The process is known as photosynthesis. Some bacteria are autotrophic, using energy obtained by oxidising inorganic substances. Nitrifying Bacteria in the soil are examples.

Autotrophic Cells. Cells able to synthesize macromolecules from simple nutrient molecules like carbon dioxide, ammonia and water.

Autotrophic Plants. The plants getting their foods exclusively from inorganic materials occurring in soil, air and water.

Autumn Fly. A non-biting fly which irritates cattle in large numbers and feeds on nose and eye secretions and open wounds. They cause cattle to huddle together and cease feeding.

Auxins. A group of plant growth-regulating substances or hormones, which are produced by growing tips of stems are roots. Artificial or synthetic auxins are also produced which are important in agriculture and horticulture, *e.g.*, inhibiting sprouting of potato tubers and thus lengthening the storage period, and preventing fruit drop in orchards. Some synthetic auxins exhibit differential toxicity, 2, 4-D is toxic to dicotyledons, but not to monocotyledons, and is used successfully to control weeds in cereal crops and lawns.

Auxotroph. Strain of a microorganism which needs some nutrient not needed by the original strain from which it was derived. An auxotroph may be produced by the occurrence of one or more mutations in a prototroph.

Available Nutrients. Chemical elements or compounds in the soil that can be readily absorbed and assimilated by growing plants ('available' should not be confused with 'exchangeable').

Available Phosphoric Acid. Refers to the sum of the water-soluble phosphoric acid (monocalcium phosphate) and citrate-soluble phosphoric acid (dicalcium phosphate) in the soil.

Available Water. The portion of water in the soil that can be readily absorbed by plant roots. Most soil scientists consider it to be the water held in the soil against a pressure of up to about 15 bars.

Available Water Supply. Refers to that amount of water which may be expressed either in terms of a rate of flow or as a volumetric quantity, and exists in a source of water supply like a stream, or reservoir, over and above the quantity necessary to supply valid and prior rights and demands.

Average Infiltration Rate. Obtained by dividing the mass infiltration that takes place during the time rainfall intensity exceeds the infiltration capacity, by the time during which infiltration takes place at capacity rate.

Avain Tubercle Bacillus. Micro-organism causing tuberculosis in swine and polutry but rarely in sheep or cattle.

Avian Tuberculin. Used for diagnosing the avian type of tuberculosis in farm animals, particularly swin and poultry.

Avidin. Protein material which can undergo combination with biotine making the vitamin to be unavailable to the body. Cooking renders avidin inactive.

Avitaminosis. Refers to vitamin deficiency.

Awn. Refers to a small bristle at the tip of a flower or fruit, especially in the grasses, or of a leaf. the wan gets attached to the husk and encloses the grains.

Awner. The part of a threshing machine which cuts off the awns, consisting of knives rotating on a fast-moving spindle.

Axenic Culture. Refers to the growing of an organism of a single species of strain *e.g.*, bacterium in a medium free of other living organisms *i.e.*, a pure culture.

Axial Gradient. Refers to the gradation of the rate of metabolism along the length of a principle axis of an animal.

Axil. Refers to the angle between the upper side of a leaf and the stem on which it is borne; the normal position for lateral buds.

Axillary . Growing from an axil. Thus an axillary bud grows from the axil between stem and leaf.

Axillary Bud. Bud greating in the axil of a leaf.

Axle-tree. The beam on a cart or wagon connecting two opposite wheels.

Azoic. Means without life applicable to precambrian era or rocks.

Azonal Soils. Refers to broad soil order in genetic classification of soils. It includes soils that have poorly developed profiles. Examples of azonal soils include sand dunes, loess, recent deposits of alluvium, and partially weathered bedrock.

Azotobacter. *Azotobacter chroocoaum.* It is a non-symbiotic nitrogen fixing bacterium occurring in the soil. It is utilizing soluble carbohydrates; strongly aerobic, susceptible to phosphate deficiency.

Azoturia. Specific disease of horse, which is caused due to excessive lactic acid, one of the byproducts resulting from the metabolism of glycogen.

B

B-cells. Beta cells of islets of Langerhans.

B-Complex. Group of accessory food factors (vitamins), which include thiamine or vitamin B_1; riboflavin B_2; pantothenic acid B_3; niacin or P.P. factor : pyridoxin B_6; biotin (H); inositol; choline; para-amino benzoic acid and folic acid etc.

B-horizon. Subsurface layer in which certain leached substances (*e.g.* iron) get deposited.

B-Line. (1) The term used for the fertile counterpart of the A-line. The B-line is not having fertility-restoring genes and is used as the male parent to maintain the A-line *i.e.* A-line *x* B-line reproduces the A-line (2) Maintainer of the A-line.

B-Trials. The term used for a series of microplot trials for ascertaining the value of a new variety over the area of its probable greatest usefulness.

$B_1B_2B_3$. Refers to first second and third backcross generations respectively. The first backcross is obtained by crossing hybrid with one of its parents, the B_2 is obtained by crossing B_1 plants with the same parent, and so on in B_3 and subsequent backcross generations. Synonyms BC_1, BC_2, BC_3.

Babcock Test. A test which is used for determining the butterfat content of milk and cream, involving the use of sulphuric acid to break down the proteins which surround the fat globules in the milk, releasing the fat for measurement.

Babiana (Baboon-root). Dwaif half-hardy bulbs with brilliantly-coloured flowers of great beauty and pale-green Lairy" leaves.

Baby Beef. The meat from cattle sold after fattening between the ages of 12 and 18 months.

Bach (Seed). The term used for the quantity of harvested crop kept into a container or a bin on a repetitive basis, specifically for treatment such as drying.

Bacillary White Diarrhoea. An often fatal bacterial disease of young chicks caused by *Salmonella pullorum*, which causes serious losses between 4 and 14 days of age. It is inherited from parent stock which themselves have survived an outbreak. Carrier parents can be detected by a simple blood test. Known as B.W.D. for short. Also called pullorum disease.

Bacillus. A genus of spore-producing bacteria. Also a general term for rod shaped bacteria.

Bacitracin. An antibiotic which has been found to be effective against bacterial infections.

Back Band. A strap, rope or chain passing over a cart saddle, and holding up the shafts of the cart.

Back Board. A board at the back of a cart.

Back Chain. A chain fastened across the back of a draught horse to the ridger rods on the shafts of a cart or wagon. Also called a ridger.

Back Country. Rural areas which are not thickly populated.

Back End. The later part of a season; the late autumn.

Back Fire. In forestry, a fire which is started intentionally ahead of and advancing fire for removing inflammable material by controlled burning and thus stop or control the main fire.

Back Furrow. Refers to a raised ridge which is left at centre of the strip of land when ploughing is started from the centare to side.

Backcross. In plant breeding it refers to a cross of a hybrid *i.e.* F_1 with one of its parents. It is carried out to transfer a specific gene from an undesirable variety to another commercially desirable one lacking in that particular character. In genetics, it refers to a cross of hybrid with a homozygous, recessive parent and the aim is to test the gametic ratio of F_1.

Backcross Breeding. Refers to the breeding method in which the backcrossing is done for several generations followed by subsequent selection until a desired combination of traits from the recurrent parent and one or two desired traits from the non-recurrent parent enters the progeny.

Backwood. Remote, uncleared forest land.

Bacon. The meat from the back and sides of a pig, preserved by Curing with brine. Bacon may be sold 'green' or after wood-smoking which produces strong-flavoured 'smoked bacon'.

Bacon Pig, Baconer. See Pig.

Bacon Side. A half carcase of a Bacon Pig.

Bacterae Mia. Refers to a condition in which viable bacteria are present in the blood. It is also called **Bacteremia**.

Bacteria. Minute, single-celled organisms which cannot be seen by the naked eye. Most are rod-shaped from *c*. 0.0005-0.005 mm long (1/50 000-1/5000 in). They are commonly called 'germs' or microbes' and inhabit soil, water, air and other living organisms in millions. Some are responsible for diseases, *e.g.* Brucellosis in cattle. Others are essential for maintaining soil fertility and aid in the decomposition or Organic Matter. Amongst the most important are those capable of Nitrogen Fixation in the soil and root nodules of certain plants.

Bacteria Count. Means the number and kind of bacteria per unit of volume present in a substance.

Bacteria Denitrificans. Denitrifying bacteria that are living side by side with the azatobacter and clostridium. They reduce nitrogen compounds in the soil and liberate nitrogen in the free state; they thrive under anaerobic conditions and may be active in soils as well as lower compact layers of manure pits.

Bacterial Ooze. Discharge, exudate of sap or gum etc. from a plant because of bacterial disease.

Bacterial Ring Rot. An important disease of potatoes especially in Canada and U.S.A. caused by *Corynebacterium sepedonicum*. Symptoms vary—usually the first sign is wilting of the lower leaves, occasionally the top leaves, followed by leaves gradually turning brown. Infected stems contain a creamy fluid. Infected tubers cut across at the stem end usually show a pale yellow vascular ring which later becomes rotten. The tuber flesh outside the ring typically easily separates from the inner flesh. Skin discoloration and cracking sometimes occur and soft rots subsequently develop, masking the ring rot symptoms. The bacteria is not soil borne, but is often spread from tuber to tuber in harvesting, dressing, storage and planting.

Bactericide. A substance which is capable of killing Bacteria.

Bactericidin or Bacteriolysin. Substance present in the sera of mammals which is capable of inhibiting the growth of bacteria, especially the gram positive ones, or of destroying them.

Bacteriochlorophyll. The main photosynthetic pigment of most photosynthetic bacteria.

Bacteriods. Irregular, enlarged forms of rodshaped bacteria which are especially of the rootnodule forming specis. These are ultimately absorbed by the cells of the rootnodule.

Bacteriology. Refers to the study of bacteria.

Bacteriolysin. Substance or agent which is able to cause rupture of bacteria.

Bacteriolysis. The destruction of a bacterium by an Antibody.

Bacteriophage. A Virus that parasitises Bacteria. Also known as a phage.

Bacteriostat. An agent that only checks the growth of bacteria but does not kill them.

Badland. A type of land generally devoid of vegetation and broken as a result of serious erosion.

Bag. The udder of a cow.

Bag Closer. Machine which is used for sewing the filled seed bags.

Bag Muck. An old term for Artificial Fertilizers, due to the fact that they are applied from a bag.

Bag Up. A descriptive term for a cow with an enlarged udder prior to the birth of a calf, due to the development of milk vessels.

Bagasse. Mill residues which are obtained from the cane-sugar industry. It consists of the crushed stalks from which the juice has been extracted. The word is also applicable to similar residues from other plants, such as sorghum, beet, or sisal, but it usually refers to sugarcane bagasse when unmodified.

Bagger Weigher. Small machines fixed beneath a bin which fill and weigh the seed bag simultaneously in a single operation.

Bagging Hook. A sickle-like tool having a square tip and smooth cutting edge. It is also known as fagging-hook, bag-hook, badging-hook. Used for the cutting of corn, pulse crops or grass. (See Sack)

Bail. 1. A division between the stalls of a stable. 2. An Australian term for a frame holding a cow's head during milking. 3. A small mobile shed for milking cows in the fields, mounted on skids for towing by tractor and divided into 6-8 stalls of a walk-through type.

Bail Milking. The milking of cows in a portable shed or Bail, either moved from field to field frequently or kept in a relatively permanent and sheltered position. Sometimes the cattle are kept in temporary or permanent yards during the winter and milked in an adjacent bail.

Bailiff. A person employed to look after or manage a farm on behalf of the owner, receiving either a salary or share of the profits.

Bait. A mixture of hay, oats, chaff and other dry food used as horse feed.

Bakers. Potatoes sold for human consumption, which are of Ware Potato standard, and which will not pass through a 65 mm horizontal mesh when manipulated, but will pass through a 90 mm mesh. (Mids)

Balance plough. An old type of one-way plough with the right and left-and left-hand bodies attached to separate beams and handles, the beams being connected together at an angle, and having a common axle and pair of wheels. While one plough operated the other was carried more or less vertically in front of it. At the end of a furrow the working beam was raised and the other lowered. Also called a tipping plough.

Balanced Daily Ration. A ration in which the nutrient ingredients are correctly proportioned for the needs of the type of stock eating it. (Maintenance Ration, Protein Equivalent, Starch Equivalent, Nutritive Ration) for 24 hours.

Balanced Feed. Feed containing all the known required nutrients in proper amount and proportion. It is prepared by the recommendations of recognised authorities in the field of animal nutrition. The species for which it is intended and conditions of growth and development should be indicated.

Balanced Fertilizer. Soil additive having suitable proportions of each necessary mineral element to develop a plant or a crop.

Bald Faced. Having white on the face, as a horse.

Bale. 1. A compressed package of hay or straw for easy handling and storage, tied usually with wire, sisal or plastic twine. Bales may be square, rectangular, round or flat depending on the type and normal operating density of the Baler used. 2. A portable field milking shed. Also called Bail or Hosier bale.

Bale Loader. A device which is attached to the front loader arms of a tractor to lift or clamp groups of Bales for loading onto a trailer or stack, or into a barn. Commonly employed in conjunction with a bale sledge drawn behind a baler. Also called Bale Slave.

Bale Slave. See Bale Loader.

Bale Sledge. A sledge towed behind the delivery point of a ram Baler, on which the operator stacks bales (usually 6 or 8), although many are now unmanned. These stacks are removed from the sledge periodically. This process allows bales to be grouped rather than scattered indiscriminately about a field by the baler, and facilitates loading and transport to the barn.

Baler. A machine that picks up Hay or Straw from a Swath or Windrow, compressing and typing it into Bales. The density to which the bales are compressed can be adjusted in accordance to the crop condition or other requirements. The most popular balers produce rectangular bales between 45 and 90 cm long, but normally 35 × 45 cm in cross section, and weighing 9-18 kg. Balers producing large 500 kg bales are becoming increasingly popular when large quantities of straw need handling.

Baler Twine. Twisted cord which is used in a Baler to tie bales. Nowadays polypropylene twine has largely replaced sisal and wire.

Ball. A Bolus for a horse.

Ball Planting. Refers to a method of planting young plants in which plants are removed from the ground with a compact ball or lump of earth round their roots for providing nourishment and help their growth in a new location.

Balling. Measuring of sugar percentage in sample syrup at fixed temperature. Also called brix.

Balling Guns, Balling Pistols. Metal instruments which are used for the administration of medicated capsules or large tablets to horses, swine and sheep.

Ballistic. Fruits with explosive dehiscence and discharge of seeds.

Balsam. A pretty half-landy annual for sunny beds or borders with rich soil or for the green-house.

Bambusa (Bambuo). There are three great classes of bamboos. Must species do will in the open, specially if the soil is of moist, deep, tight, loamy nature and has some heat or lead-mould in it.

Banana Flour. Also known as banana meal. Meal prepared from fully mature unripe bananas.

Band Placement. Refers to the fertilizer placement in bands on one side or both the sides of the row, about two inches below the seed and two inches to one side of the seed or plant grown.

Band Seeding. The term used for placing the seeds in rows directly above, but not in contact with a band of fertilizer.

Banding. 1. The tying of bands of grease round fruit trees to trap insects. 2. The fastening of corn in Sheaves.

Banding-in. The tying together, at breast height, of the several Bines arising from a hop plant.

Bandster. A person who binds sheaves of corn after the reapers (obsolete).

Bantam. A small variety of domestic fowl with feathered legs. Said to originate from the town of Bantam in Java.

Bar Eye. Name assigned to one of the mutations of Drosophila in which the number of optical units in the compound eye has been reduced.

Bar Pig. A hog; a castrated male pig. Also called barrow pig.

Barb. 1. A swift breed of horse. 2. A dark-coloured fancy pigeon.

Barban. A translocated herbicide which is used either alone or in mixtures to control Wild Oats and other grass and broad-leaved weeds.

Bare Fallow. Land, particularly heavy land, left fallow for a year, during which it is ploughed several times so that the sun and wind dry the soil during the summer months, desiccating and killing perennial weeds. Success depends on summer rainfall. Bare following is used less frequently nowadays due to the availability of herbicides.

Bare Land Holding. Farm land lacking the basic buildings necessary for agricultural production.

Barfored's Test. Biochemical test which is used for testing monosaccharides. Barfored's reagent is an aqueous solution of copper acetate and acetic acid. It is added to a solution or suspension of the suspected monosaccharide. The positive result is a red precipitate; the negative result is the absence of any precipitate. In this test there occurs the reduction of cupric accetate to red insoluble cuprous oxide.

Bark. 1. Refers to the cork and other dead tissues to the outside of the functioning phellogen. 2. In medicine and horticulture, it refers to all the tissues outside the cambium and so includes the phloem.

Bark Miner. An insect the larva of which spends most of its developmental period in the outer corky bark.

Barley. A cereal probably the first to be cultivated by man, grown for its grain which is used either as stock feed or for beer making. The bearded seeds in the ear are arranged in 2, 4, or 6 rows.

Barleycorn. The grain of barley from which malt is made (personified as John Barleycorn). Also a term for a single grain of barley.

Barley Loose Smut. Fungal disease of barley which is caused by *Ustilago nuda*. It is charaterized by smutted heads where grains are replaced by black powdery mass of spores, leaving the rachis later naked.

Barley Water. Squash of fruit in which part of added water is replaced by an extract of dehusked and polished pearly barley.

Barium Carbonate.White, water insoluble powder which is used as rodenticide.

Barm. Yeast. Also the froth of fermenting liquor.

Barn. A building in which grain, hay and implements, etc, are stored.

Barn Allergy. An allergy, similar to hay fever, the symptoms of which include a runny nose, sneezing, inflammation of the eyes or more severe symptoms. Unlike farmer's lung, with which it is associated,barn allergy symptoms occur at the time of contact with dust from hay or grain. The principal cause is the presence in stored feedingstuffs, such as hay and grain, of large numbers of mites which multiply soon after baling, often producing a pinkish dust on the barn floor. Up to 4 million mites may be present in a single bale of hay. The most effective solution to the allergy is the use of a respirator.

Barn-Cured. Cured within a barn, as contrasted to sun-or air-cured.

Barn Machinery. A general term for implements or machines which are not used in the fields, *e.g.* Hammer Mill, Grain Drier, weighing machines, chaff cutters, etc.

Barnevelder. A heavy breed of fowl, with black and white varieties, although similar in colour to a partridge. It has a single comb, yellow legs, and the hens produce dark brown eggs.

Barnyard. The area adjoining a barn. The farmyard.

Barrage Sexual. Repulsion between sexually incompatible hyphae. Incompatible strains will grow away from each other.

Barred Feathers. Polutry feathers exhibiting a horizontal pattern of black and white stripes or bars. Other colours may be combined with black such as buff or gold, thus giving buff-barred and gold barred feathers.

Barrel Bulk. A measurement of five cubic feet.

Barren.(*a*) (Of animals) incapable of producing offspring, due to infertility or sterility. (*b*) (Of plants) not producing fruit or seed (*c*) (Of land) bare of vegetation, incapable of producing useful crops; arid.

Barrener. A female farm animal adjudged to be baren. A term usually applied to cows.

Barrier. Any type of obstruction-physical, chemical or biological that does not allow the migration of animals, or gradual extension of their territories.

Barbon. A farm yard.

Basal Area. Refers to the area of the cross-section of a stem. It is usually 6 ft. above union of a tree at breast height. When it is applied to a crop ("Crap basal area"). The sum of basal areas of all the stems or the total basal area per unit of area. It may be expressed in true or in quarter girth units.

Basal Doso. Manures or fertilizers which are applied to the soil before the crop is sown or planted or transplanted.

Base Exchange. Process in which soil absorbs certain positively charged ions from the soil solution and releases other cations in equivalent quantities.

Base Exchange Capacity. Amount of base which is held on a clay under specified conditions *e.g.*, of pH.

Base Saturation. Extent to which a material is saturated with exchangeable cations other than hydrogen, expressed as a percentage of the cation exchange capacity.

Base Saturation Percentage. The extent to which the Adsorption Complex of a soil is saturated with exchangeable Cations other than Hydrogen and Aluminium. It is expressed as a percentage of the total Cation Exchange Capacity.

Base Status (High or low) Quantity of nutrient bases available in the soil.

Basic Cleaning. Means cleaning of the seed by separating material smaller as well as larger than the good seed, general size grading, and cleaning. For example, cleaning by Air Screen Cleaner.

Basic Slag. A by-product of steel manufacture which is used as a Fertilizer. It is a fine, powdery, grey-black, and rich in Phosphate and lime.

Basidiomycetes. A large group of Fungi, having most of the familiar types like mushrooms, toadstools, bracket and jelly fungi and puff balls, and also parasitic forms such as Rusts and smuts. Spores are produced by club-shaped cells called basidia.

Basidiospore. Sexually-derived spores which are produced following the union of two nuclei on a specialized clublike structure called a basidium.

Basil. 1. An aromatic herb which is used in cooking. 2. A sheepskin, which is roughly tanned and undressed.

Basket Brooder. Brooder which is made with an ordinary domeshaped basket, lower half of which gets lined with guuny sack or cement mixture wash and upper half left open for allowing passage to smoke hurricane lantern source of heat, sleeved off by ½ inch wire mesh; floor has been covered with straw or saw dust.

Bast. Phloem. The inner bark of a tree which is used in long strips for basket-and mat-making, and for tying plants to stakes.

Bastard Fallow. Land left fallow for half the summer, between the harvesting of one crop (usually silage or hay) and the sowing of the next crop, during which it is ploughed to kill perennial weeds by desiccation. The procedure remains the same as for a bare-fallow except that it does not waste a year. Success is very dependent on favourable weather conditions. Also known as half allow, pin fallow short fallow or summer fallow.

Bastard Trenching. A method of deep digging in which the soil layers get replaced in their natural sequence, as opposed to ordinary trenching in which the sequence of soil layers taken with each spit get reversed when replaced.

Bat. A pole which is used in hop gardens for supporting wire-work usually Sweet Chestnut or Larch.

Batch Drying. A method of drying Bales of Hay in batches in a barn, usually using heated air blown through a ventilated floor. Batches of about four layers have been dried at a time until moisture content gets reduced below 20%, then removed for storage, and the next batch introduced. This method has been land to be less popular nowadays than storage drying and tunnel drying.

Batesian Mimicry. Refers to the resemblance of a harmless animal to a poisonous or damagorous one providing protection to the former because predators will tend to avoid both. The clearwing both resembling a bee is a well-known example.

Batfowling. Catching birds at night by showing a light and beating the bushes.

Bating. One of the steps which are used in the tanning of leather. It is following dehacring with lime. In this the skins are soaked in a bath containing enzymes and ammonium salts. The aim is to remove residual lime, hair, skin, and glands.

Batology. The study of brambles.

Batt. 1. The long wooden handle of a Scythe. 2. A faggot of stout sticks.

Battery Hens. Hens kept for their laying lives indoors in Cages, normally in three-tiered rows, Laying batteries may be static, semi-or fully-automatic, depending on the method of feeding, watering and manure clearance. Battery housed birds achieve a higher rate of egg production than those under other systems. More than 96% of the national laying flock in Britain is now housed in battery cages.

Baule Unit. Baule postulated that the unit of fertilizer of any other growth factor, may be taken as that amount which is necessary to produce a yield that is 50 per cent of the maximum possible. According to this concept one Baule unit of a growth factor is equivalent to one Baule of any ohter growth factor in terms of growth-promoting ability. The values of the Baule unit in pounds per acre of N, P_2O_5 and K_2O have been 223.45 and 76 respectively based on Mistcherlich's work.

Baulk. 1. An unploughed strip of land left to mark the boundary between fields. 2. A ridge created by ploughing and then left unploughed. 3. A beam of a timber. 4. A ridge (balk) in which potatoes are planted.

Baulking Plough. See Ridger.

Bay. 1. The space between two columns in a barn, or any recess in a farm building. 2. A stall in a stable. 3. The combined cry of hounds hunting an animals, or the last stand of an animal cornered by hounds. 4. A description of a horse, reddish brown to chestnut in colour.

B.C.P. Test. A test which is used to determine the presence of antibiotics in milk (Milk Quality Schemes) which superseded the T.T.C. test in 1976. It involves the addition of a bacterial culture to a milk sample together with the colour indicator Bromo-Cresol-Purple (B.C.P.). If no antibiotics are present the culture grows, produces Lactic Acid, and causes the B.C.P. to change colour through green to yellow. If antibiotics are present no growth takes place no acid is produced, and there occurs no colour change.

Beam. 1. A large straight piece of iron or timber which is usually square in cross-section and is forming one of the main structural members of a building, normally supporting rafters. 2. The main shaft of a plough, wooden or metal, to which are fixed the Mouldboard, Coulters and wheels, and the handles of horse drawn ploughs. This part transmits the power of the animal to the plough. 3. The main trunk of a stag's horn.

Beamage. An allowance on an animal carcase for loss of weight by evaporation.

Bean (s). Various leguminous plants, and their seeds borne in pods, distinguished from vetch, clover and peas by having a square, hollow steam. They are tall, errect plants, varying in length according to variety, and develop a strong tap root bearing nodules containing bacteria capable of nitrogen Fixation. When ploughed-in, valuable nitrogen and organic Matter is returned to the soil. The seeds or beans are rich in Protein.

Bean Straw. The Haulm of bean plants.

Beard. 1. The hairy tuft on the lower part of a goat's jaw, or on a turkey's neck (also called Brush). 2. A term for an Awn or threadlike spike, as on the ears of barley

Beaters. 1. Steel bars on the drum of a threshing machine of combine harvester which dislodge the grain out of the ears of cereals as they rotate. Also known as beater bars. 2. Those who rouse game in shooting or hunting.

Beaumont Period. A period of 48 hours during which temperatures do not fall below 10°C (50°F) and relative humidity remains above 75%. Outbreaks of Potato Blight may be expected within 3 weeks of such a period.

Beck. 1. A brook or small stream in the north of England. Equivalent to a Burn in Scotland. 2. A type of hand-hoe used to chop the soil around a hop plant.

Bed. 1. A garden plot. (Seed Bed) 2. A layer or stratum of rock. 3. A sleeping place for an animal. (Bedding)

Bed Bug. *Cimex lectularius* or *Cimex hemipterus.* A blood sucking insect which belongs to the genus Cimex. It lives and lays its eggs in the crevices of bed steads, furniture and walls.

Bedded Set. A young hop plant, having rooted in a nursery bed from a cutting.

Bedding. Litter for farm animals to sleep on, usually straw, shavings, sawdust, etc.

Bedding Land. The ploughing, grading or otherwise elevating the surface of fields into a series of parallel beds or 'lands' with shallow surface drains separating them.

Bedding Soil. Arranging the surface of fields by ploughing and grading into a series of elevated beds separated by shallow ditches for drainage.

Bedrock. The solid rock beneath the soil.

Bee. A furry insect of the Order Hymenoptera. There are solitary and social bees. The latter live in colonies. Each colony has one queen bee together with large numbers of female 'workers' and a fewer male 'drones'. The most highly socialised is the honey-bee kept in hives by beekeepers for commercial Honey production. Bumblebees have much smaller colonies. A growing practice is the hiring and placing of Beehives in Orchards to promote pollination and fruit setting.

Bee Beard. 1. The pollen of flowers collected by bees and fed, together with honey, to the larvae. 2. The local name of several plants yielding nectar.

Bee Brush. Brush or whisk broom often used to brush off bees from a honey-comb before it is taken away for extraction.

Bee Candy. Artificial food for bees which is made from sugar and substances like honey.

Bee, Carpenter. A member of the family Oxylocopidae. It makes tunnels in dry wood or twigs for nesting purposes.

Bee Escape. A device which allows bees to go through a self-closing exit in a beehive.

Bee Farming. The term used for the keeping of bees for honey and beeswax they produce, and for their pollinating of planars.

Bee Fold. An enclosure for beehives.

Bee Hole. Commonly applied to the hole which is bored in teak by *Xyleutes ceramicus,* (the beehole borer).

Bee Parturage. Also bee forage. Plants yielding necter and pollen, and within flying distance of a hive.

Bee Smoke. A device having a funnel, firepot and bellows for blowing smoke into a beehive to force out the bee colony.

Beef. Meet which is derived from cattle nearly one year old or older.

Beef Bull. A bull which is reared and fattened to produce beef. Male cattle for fattening are usually castrated to produce steers. Bull carcases are leaner than those of steers, with darker muscle and a heavier forequarter. The advantage of beef bull production is that bulls grow faster, convert food more efficiently, and achieve greater carcase weight than steers.

Beef Cows. Cow and hiefers which are kept mainly for rearing calves for beef production, as distinct from dairy cows.

Beef Shorthorn. A breed of compact, short-legged beef cattle.

Beehive. A box in which bees are kept to produce honey. Hives contain two separate chambers, a lower brood chamber in which the queen lays and rears her brood, above which is the honey chamber or supers containing frames in which the bees store surplus honey which is periodically removed.

Beeswax. Wax secreted by bees and used to construct the cells of the honeycomb. Used in polishes and cosmetics.

Beet. 1. A plant (*Beta spp.*) of the goosefoot family with a succulent root, used as a food and as a source of sugar. 2. A sheaf of harvested flax.

Beet Cyst Nematode. A cyst-forming, plant-parasitic eelworm or rounderworm (*Heterodera schactil*) which attacks beet.

Beetle. 1. An insect of the Order Coleoptera, in which the forewings are reduced to hard and horny protective covers for the hind-wings and body. 2. A heavy wooden hammer with a long handle used for driving wedges and posts, or crushing paving stones, etc.

Beeves. Cattle.

Bettle, Death Watch. *Xestobium rugovillosum* Deg. It is the species of the family Anobidae. It breeds in dry wood, particularly in old timbers in buildings, and reduces it to a honey-comb like structure.

Beetle Engraver. Certain members of family Scolytidae, whose bark tunnels form characteristic patterns having one or more egg galleries issuing from central nuptial chamber.

Beetle Longhorn. It is a member of the family Ceramycidae. Its larva bores in the wood of cambium mainly of injured, fallen, dying or dead trees.

Beetle Power-post. A member of the families Bostrychidae or lyctidae, It bores mainly in sapwood of seasoned timber and leavidg a fine flour-like dust.

Belgian Blue. A breed of cattle found in Belgium, South Netherlands and Luxembourg.

Belgain Hare. A breed of rabbit with the appearance of a hare, having an abnormally long head, ears and legs. Rich red or chestnut in colour with dark, wavy ticking.

Bell. 1. To bellow or roar. The cry of a stag at rutting time. 2. The catkin containing the female flowers of the hop. Also to be "in flower.

Bell Wether. The leading sheep of a flock, with a bell on its neck.

Belly Band. A saddle-girth. Also a strap fastened to the shafts of a cart and passing under the belly of a horse towing it, so that the cart is prevented from tipping backwards.

Belt. 1. A zone of country. 2. A term for sheep scab.

Belt Transect. A narrow strip of land taken as a sample of an area in studying the vegetation.

Belted Galloway. A variety of Galloway cattle with a white belt round the body. Also called Beltie.

Beltie. Belted Oalloway.

Benazolin. A translocated herbicide, used only in mixtures, for controlling broad-leaved weeds.

Bench Cross-Section. Refers to the cross-section for contour benches. It is an important phase in planning for irrigation. Width of strip between ridges is kept such that it fits the farm equipment to be used 'height of the bend is kept sufficient to contour both normal irrigation stream and storm runoff.

Bench Terrace. It is the commonest type of terracing which is followed in India. It is constructed to make sloping land cultivable or stable. It has a series of platforms or mostly level benches which are cut into the slope in a step sides often with retaining walls supported by rock or vegetation. Vertical drop 2 to 6 ft. depending on slope and soil conditions. Small shoulder bund (1 ft. height) is also constructed along outer edge of terrace.

Benedict Solution. An alkaline copper sulfate solution, blue in colour. When a few drops of a reducing suger solution are added after heating the copper solution, the cupric copper gets reduced to the cuprous state and a change to a reddish colour is observed. It finds use as a test for the presence of sugars in solution.

Benedict's Test. A biochemical test whcih used for testing sugars. The Benedict's solution is prepared by mixing an alkaline copper sulphate solution with sodium citrate. The positive result for relatively large quantities of sugar and the negative results have been the same as those of Fehling's test.

Beneficial Cultivations. Those cultivations which enhance the condition of the land, or are good for crops.

Benomyl. A systemic fungicide.

Bent. A general terms for short flowering stems of grasses projecting above the turf and old dried grass-stalcks.

Benthic. Animals inhabiting the bottom of the sea (or fresh water.)

Benthos. A term used for organisms that are living collectively along the bottom of oceans and lakes.

Benzene Hexachloride. Also called BHC. It is an effective and economic insecticide available as dusting powder, water dispersible powder, emulsifiable concentrate or solution and as fumigant in highly volatile solvents. It has 5 isomers of which gamma has been most important due to insecticidal value; insects killed by contact or by stomach of fumigant action, can be used in mixture with other insecticides and fungicides, except those having strongly alkaline materials like Bordeaux mixture, lime or lime sulphur.

Beri-beri. An illness by an acute deficiency of vitamin B (thiamine).

Betaine. A colourless, tasteless, crystalline substance present in sugar beet, producing a strong fishy taste on decomposition.

B.H.C. Benzene hexachloride, a persistent organochlorine insecticide used mainly in sprays and dusts. Particularly used to control wireworms and other soil pests.

Bible. The third stomach of a ruminant.

Biddy. A Hen.

Bident. A two-year-old sheep.

Biennial. A plant that flowers and bears fruit only in its second year, then dies. Examples include carrot and beetroot.

Bifoliate. Having two leaves or leaflets.

Big Bud. A swelling of current buds due to a gall-mite.

Bigg. Four rowed barley. Also called bear.

Big Leg. An animal disease which gets developed either in the front or hand legs and may be extending to the shoulder, neck and rump. It occurs due to vitamin A deficiency. The affected animals lose weight rapidy and move about slowly.

Big Neck. A goiter (of animals) which occurs due to iodine deficiency.

Bighead. An animal disease which is affecting ewes, lambs and goats during summer. It has been characterized by swelling of the parts of the head that are bare of lightly covered with wool. It occurs due to a toxic condition due to horsebrush poisoning (affecting the liver) and a swelling affecting the head. The disease appears in a band 15-25 hours after the animals have eaten the poisonous plants. The ears, eyelids, face or lips start swelling. The swollen parts of the head become hot and painful. Treatment involves confining the affected animal to a darkened barn feeding dry, laxiative feed such as barn, legume hay and plenty of fresh and clean water. Seriously affected cases need more care. Bathing the affected parts with an epsom salt solution is also effective. The disease could be prevented by confining the flock in a shed, barn or some other shaded area during midday.

Bile. A bitter, yellowish green secretion of the liver, important in the digestion of fats, and stored in the gall bladder. Also called gall.

Biliary Pigments. Pigments formed in the liver from the haemoglobin of old red cells that are destroyed there.

Bilirubin. One of the bile pigments.

Bill Hook. A hatchet with a curved point, used in hedge trimming, etc.

Billy Goat. A male goat. Also known as a buck.

Bin. 1. A large receptacle for storing feedingstuffs, corn, fruit, etc. 2. A canvas container on a collapsible wooden frame used initially for storing hops when picked. Also called a crib.

Bin Sampler. Large size trier constructed on the principles of big triers for drawing samples from bins.

Binder. A machine (full name: reaper-binder) which cuts and binds corn in sheaves. Powered by p.t.o., although older and heavier machines obtained power from a large bull wheel. It produces long, unbroken straw, used in thatching.

Bing. A feed passage, particularly for sheep.

Bink. A wasps' or bees' nest.

Bio-Assay. Refers to the evaluation of the effect of a substance on an organism under controlled conditions *e.g.*, the bio-assay of fungicides.

Biochemical. Deals with the study of chemical substances occuring in living organisms and the reactions and methods for identifying these substances.

Biochemistry. Deals with the study of chemical processes and substances occurring in living cells.

Bioengineering. Deals with the manufacture and use of artificial replacements for body organs that have been lost or are malfunctioning, like artificial limbs, heart pacemakers.

Biogas. A mixture of methane (70%) and carbon dioxide produced as an alternative source of energy by the anaerobic digestion of waster material–mainly animal slurries, but also crop residues and industrial wastes. Research continues to perfect the technique and includes the use of digested slurries as animal feedingstuffs.

Biological Erosion. Erosion of soil by water or wind because of the soil being exposed by the burrowing of rodents, destruction of vegetation by insects etc.

Biological Clock. Physiological mechanism in plants and animals which makes them to have a knowledge of time and to take account of it, even if the normal indicator of day and night luminosity are lacking.

Biological Control. Refers to the reduction or control of a pest by introducing a suitable predator into its habitat *e.g.*, the control of the greenhouse whitefly Thriaeurodes by the minute chalcid wasp Encarsia.

Biological Interchange. The term used for the interchange of elements between organic and inorganic states in a soil or other substrate through the agency of biological activity. It occurs due to the biological decomposition of organic compounds and the liberation of inorganic materials on the one hand (mineralization), and the utilization of inorganic materials in synthesis of microbial tissue on the other (immobilization). Both processes commonly take place continually in normal soils.

Biological Mineralization. The conversion of an element occurring in organic compounds to the inorganic form because of biological decomposition.

Biological Weathering. Physical and chemical weatherings which are assisted by biological agencies.

Biomass. Refers to the total weight of living matter in a population. It is usually expressed in terms of dry weight per unit area.

Biopsy. Diagnostic examination of a piece of tissue which has been removed from the living organism.

Biosynthesis. Refers to the coming together of chemical building units to form new materials in living plant or animal.

Biosystematics. The science of classification of organisms.

Biota. Fauna and flora of a region.

Biotin. It refers to one of the ten recognised vitamins of the B complex. It is widely distributed in foods.

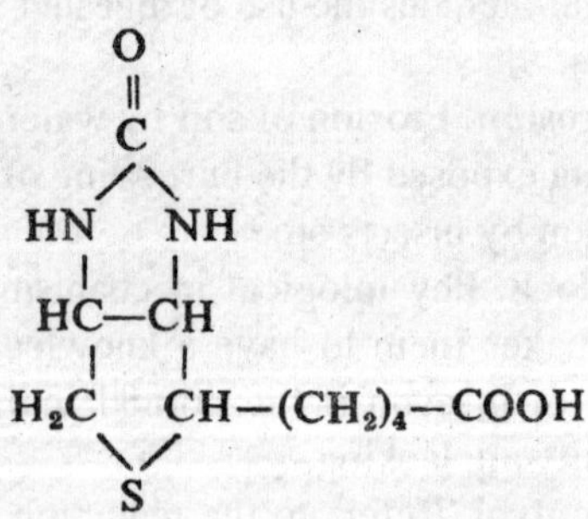

Bird Minding. Scaring birds away from field and orchards by creating noise, *e.g.*, shouting, firing guns, etc. Also called bird scaring or bird starving.

Bird Scarers. Device for making loud noises at intervals which scares away the birds.

Bite. Grazing. Early bite is grazing in the early spring. Late bite is grazing at the end of growing season.

Bitter Pit. A disease of apples, characterised by brown spots and depressions.

Biuret Reaction. It is a specific test for the detection of peptide linkage. It involves adding of alkali and a few drops (3-4 drops) of a dilute solution of copper sulphate (about 0.02%) to the solution. This produces the development of a bluish to pink colour depending on the type of protein present.

Biuret Test. *See* **Biuret Reaction.**

Black Blowfly. Develops in decaying meat.

Black Disease. An acute malady which is affecting mature sheep in good conditions. It is caused by the germ *Clostridium nouyi* in the presence of liver flukes.

Black Heart. A type of winter injury in which the inner wood gets darkened while the cambium and the bark remains alive *e.g.*, nursery trees are susceptible to this type of injury.

Black Land. Heather moors, dark in appearance. Also a term for the dark, humus-rich, fenland soils.

Black Mustard. A cruciferous plant (*Brassica nigra*) grown for its seeds to produce table mustard. Seeds also contain up to 22% oil, extractable under pressure.

Black Rust. A fungal disease (*Puccinia graminis*) of wheat, oats, barley, rye and several grasses. Forms reddish brown (becoming black) spots or lines on stems and leaf sheaths. Its alternative host is common barberry (*Berberis vulgaris*).

Black Soil. Occurs in low lying areas having 55 to 65% clay content, good supply of lime and the pH is 7-8. Nitrogen content has been medium, potash is fairly well supplied P_2O_5 is low. It is quite fertile and suitable for wheat, cotton, chillies and jowar.

Blackcurrant. A shrub (*Ribes nigrum*) grown for its small black berry.

Blackhead. A disease of turkeys caused by a protozoan, *Histomonas meleagridis,* infecting the liver and intestines. Symptoms include ruffing of feathers, loss of appetite and a mustard-yellow diarrhoea. A common cause of loss in turkeys, particularly when young.

Blackleg. 1. A bacterial disease (*Erwina atroseptica*) of potatoes which causes blackening and rotting of the stem base. Leaves lose colour, turn brown, and stems die. Brown rot infects the tuber. Principally a seed borne disease. 2. A fungal disease (*Phoma batae* and *Pythium spp.*) of sugar beet and mangold seedlings, which become black at ground level, and threadlike, then wilt and die. 3. A bacterial disease (*Clostridium spp.*) of sheep (also called gas gangrene) and cattle (also called blackquarter or felon). In sheep *C. chauwaei* infect ewes via wounds, often at lambing, dipping or shearing, and lambs at castration or docking. It causes hot painful swellings in muscles which darken and may be gassy, and death within hours. In cattle, bacteria from soil infect minute wounds, particularly in calves, producing symptoms similar to sheep.

Blade. 1. The flat or expanded part of a leaf. Cereals are 'in blade' when flat, the thin, long leaf is formed but before the corn ears have developed. 2. The thin cutting edge of a knife, axe, scythe, etc.

Blae. Hardened clay or carbonaceous shale, often blackish or dark bluish, and spread on arable land to improve soil texture.

Blanchnig. Providing conditions under which growth gets forced to take place in darkness, resulting in tissues or organs with little or no green colouring matter.

Bland Oils. Oils often used as inert, soothing ingredients in medicines, *e.g.,* olive oil, cottonseed oil, etc.

Blaney-Criddle Formula. It states that the amount of water consumptively used by crops during their growing season gets closely related with mean monthly temperature and daylight hours.

Blaze. A white mark on an animal's face.

Bleat. Call of sheep and goat.

Blemish. Includes green patches, brown spot and patches, damage which occurs due to pest and disease, breakage in handling, sponginess and black spots, as in tobacco.

Blend (1) To cross-broad plants or animals to obtain an offspring having characteristics of each (2) To mix two or more ingredients to get a standardized product.

Blending Inheritance. The inheritance of characters so that the offspring, and successive generations become intermediate between the original parents. This is because the character has been controlled by several genes.

Blet. Incipient internal decay in fruit.

Blight. A plant disease caused by fungal parasites and various insects, Aphids, Mildew, Rust, Smut, etc. Usually a blight will attack a whole crop and sometimes a particular crop throughout a region. Nowadays the term is usually restricted to Potato Blight.

Blind. A plant which fails to flower or produce fruit.

Blind Cultivation. Cultivating with a harrow weeder, rotary hoe, or other implement to kill weeds before a seeded or planted crop has come up.

Blind Eyes. Shoots which do not produce flowers.

Blind Gut. The Caecum.

Bloat. The swelling of a cow's rumen due to gas from fermentation of green food, particularly lush grass containing white clover, or from frosted or muldy food, and obstructions in the gullet. It causes respiratory distress and, in acute cases, death. Also called hoven, rumen tympany or blast.

Block. A main territorial division of a forest, which is generally bounded by natural features and bearing a local proper name.

Blood Mare. A throughbred Mare.

Blood Meal. A residue from a slaughterhouse, rich in protein, low in mineral content, used as a feedingstuff, particularly in rations for pigs and poultry.

Blood Sports. Those sports involving the killing of animals, *e.g.*, fox-hunting.

Blood Stock. Thoroughbred horses, bred through generations for their excellent qualities.

Bloodline. Succeeding generations of animals of a species which are specially bred to maintain the genetic composition responsible for specific desirable features or qualities. Maintaining a bloodline is also known as breeding true.

Bloom. Grains, short rods, or crusts of wax on the surface of some leaves and fruits.

Blossom Honey. Honey produced wholly or mainly from the nectar of blossoms.

Blossom Wilt. A fungal disease (*Sclerotinia laxa*) of apples, pears, plums and cherries, causing fruit to rot and blossom and young leaves to wilt and die.

Blotch. Disease which shows large and irregular spots on stems, shoots and leaves.

Blotch Mine. Dark patch on a leaf which is caused by a minute insect larva mining or burrowing between the upper and lower epidermis. Various insects do this, including the larvae of small moths like the Lilac leaf-miner (*Gracilaria syringella*) and of Diptera such as the Celeyfly (*Acidia heraclei*). Blotch Mines are only one of various patterns which are produced by leaf-mining insects.

Blow. 1. To develop bloom or blossom. 2. The condition of a cow with its stomach swollen by gases, mainly methane. 3. A term for soil blown by the wind.

Blow Fly. 1. A general name for a number of species of fly which deposit their eggs in flesh and carcases. Also called flesh fly. 2. The Bluebottle, a two-winged fly with metallic blue wings. The larvae feed on animal matter and dung.

Blowout (Erosion). An excavation in areas of loose soil, usually sand which is produced by wind action.

Blown. 1. A descriptive term for sheep suffering from attack by maggots, particularly of the Blow Fly. 2. A descriptive term for an animal suffering from bloat.

Blown Out Land. An area from which all or almost all the soil has been removed by wind erosion. Usually barren, and unfit for crop production.

Blue Ointment. Mild mercurial ointment having a bluish colour. It is used for the destruction of external parasites particularly for the extermination of lice.

Bluetongue. Sore **Muzzle of Sheep.**

Boarding. The practice of tilting a plough towards the ploughed land to increase the pressure on the Mouldboards.

Bobby Calf. A term for the unwanted male offspring of a milk producing breed of cattle, usually a Channel Island Breed, used for veal.

Bodge. An old measure for oats, approximately a gallon, or half a peck.

Body. The operating part of a plough comprising a coulter, share and Mouldboard. Most modern ploughs have several bodies.

Bog. In general terms it refers to a spongy, usually peaty, wet area of marshy land. The term is applied more strictly to wet, very acid Peat, characterised by *Sphagnum* moss, found particularly in areas of heavy rainfall. Distinct from a fen which has an alkaline to slightly acid peaty soil.

Bog Soils. An intrazonal group of soils having a muck of peaty surface underlain by peat, developed under swamp or marsh types of vegetation. It is mostly in humid or subhumid climate.

Bog Spavin. Accumulation of synovial fluid in and around the hock joints of horses.

Bogey. A low heavy cart used in the past for carrying Haycocks.

Bogle. A scarecrow.

Boiler. A pullet or Hen which is sold for the table at the end of the laying period.

Boll. 1. An old measure of capacity for grain. In northern England varying from 2 to 6 Bushels ; in Scotland usually 6 bushels. 2. An old measure of weight for grain equal to 140 lbs (63.5 kg).

Boll Weevil. The larva of a noctuid moth *Heliothis armigera*. It feeds on cotton bolls and other seeds.

Bolt. 1. A bundle of willow shoots or reeds about 3 ft in circumference. 2. To break away, as when a horse suddenly dashes off. 3. A sieve for separating Bran from flour. (Bolting Cloth).

Bolt, Carriage Type. Commonly used bolt in wood. It is having a round head and a square or splined neck which grows into the wood to keep the bolt from turning.

Bolter. 1. A term used of biennial plants, which normally take two years to flower and set seed, but which get out of hand and produce seed instead in the first year, thus reducing the harvest yield. Examples of crops which sometimes bolt include cabbage and sugar beet. 2. A Bolting Cloth.

Bolting. Formation of elongated stem or seed stalk. It usually takes place during the second season of the growth in biennial plants.

Bone. To remove the bones from a carcase.

Bone Ash. Produced by burning bones with free access to air and contains 15 to 16.5 percent phosphorus. It is used as a feed supplement.

Bone Chewing. Animals on feed which are low in phosphorus commonly chew bones, wood dried or decayed animal carcasses as an instinctive reaction to the dietary deficiency. Affected animals get unthrifty, and milk production reduces to a minimum. The remedy involves feeding of phosphorus rich feeds such as oil cakes.

Bone Feedstuff. The term used for a common supplementary source of phosphorus for animals. It is a natural source of calcium phosphate which is provided by bones in various forms. They supply phosphorus and calcium in animal diet.

Bone Flour. Finely ground bones, containing phosphate, used as Fertilizer and also as a Feedingstuff.

Bone Marrow. Yellowish tissue that occupies a bone cavity.

Bone Meal. Bones, coarsely ground, after removing most of the rat. Used as a feedingstuff of fertilizer. (Meat-and-Bone-Meal).

Bone Seeker. Compound or ion that migrates *in vivo* preferentially into bone.

Bonnet. The second stomach of a Ruminant.

Bonnney clabber. An Irish term for milk naturally clotted on souring.

Boost. An instrument used to mark sheep, often with the owner's initials, usually using hot tar.

Boot. 1. Upper leaf sheath of a grass. 2. Stage at which the inflorescence expends the boot.

Borcic Acid. Boric acid, white powder used in medicine as very weak antiseptic.

Bordeaux Mixture. A mixture of copper sulphate (bluestone or blue vitriol), Lime, and water, used as a fungicidal spray on crops. (See Burgundy mixture).

Border. Strip of land, which varies in width and length edged on both sides by a low ridge which guides the flow of water within that strip.

Border Irrigation. Also known border strip irrigation. It refers to system of irrigation in which the field has been divided into border strips by using low ridges; water turned into upper end of each border strip moves down the slope in a thin sheet. It is an efficient, rapid and reltively easy method of irrigation. It is useful for all close-growing crops, some row crops and orchards where topography and soils are suitable.

Border Rows. In hybrid seed plot, recommended number of rows of the male parental line grown on all the sides of the field.

Border Strip. Grassed thickly vegetated strip at the edge of a field, along outlet channels, or at ends of rows to check or prevent erosion.

Boron. (B). A chemical element present in borax and boric acid, and essential to crops, epecially roots, in trace quantities only. Excess lime can induce deficiency, particularly in light soils, and result in deficiency diseases (*e.g.* Brown Heart, Hearts Rot).

Bosk. A thicket or little wood.

Boss. Rounded or knoblike protuberance, as on the side of a bone or tumor.

Bot Fly. A term applied to several species of two-winged flies parasitic on ungulates. Examples are *Oestrus ovis* which infests sheep, and *Gastrophillus intestinus*, the Common Horse Bot, the larva of which develops in the horse's intestine.

Bottle. A bundle of Hay or Straw.

Bottom Growth. Those grasses and clovers in a pasture growing close to the ground, as opposed to the taller plants, or top growth.

Bottom Land. Alluvial deposits such as a floodplain.

Botulism. Food poisoning which occurs due to the toxin of *Clostridium botulinum*. Fungal food poisoning takes place in man and animals Fungal toxins include, aflatoxins, phallotoxins and muscarine.

Bougie. Cylindrical instrument which is used for insertion. It is used in the treatment of strictures.

Bound Stock. Those animals, which usually have been bred on a farm, such as a flock of sheep or a dairy herd, which remain with the farm when it is sold or changes hands. Also used to mean sheep acclimatised to a Hirsel.

Bovicola. Blood sucking lices.

Bovine. An animal belonging to the cattle group.

Bovine Growth Hormone. A proteinous substance produced and released by the pituitary gland which is located at the base of the brain.

Bovine Ketosis. A disease of cattle caused by an excessive drain on blood sugar for the production of Lactose, which impairs energy metabolism and results in the accumulation of toxic levels of ketones in the blood, milk and urine. Symptoms include loss of appetite, sweet-smelling breath or milk (due to the presence of acetone and acetoacetic acid), falling miik yields, loss of weight, lethargy, and nervous behaviour. Most animals recover after treatment with glucose, glucose-forming substances or glu-coneogenetic hormones. Also called acetonaemia or acidosis.

Bovine Leucosis. A cancerous disease,Leukaemia, of cattle. It may remain begin throughout life, or become malignant usually in cows 4 to 8 years old, resulting in fatality.

Bovin Malignant Catarrh. Highly infections disease which is charater-ized by fever, catarrhal inflammation of the oral, nasopharyn-geal and conjuctival mucous membranes, general lymphoid hyperplasia and frequent involvement of the central nervous system.

Bovine Tuberculosis. A chronic contagious bacterial disease *Myco-bacterim tuberculosis*.

'Boy's Land'. A colloquial term for Light Soil, easily cultivated, as distinct from man's land.

Brace. A pair of dead (shot) Game birds.

Bracken. A tall fern (*Pteridium aquiliunum*), also called Brake Fern. It is common on hillsides, heaths, moors and in woods, where burning or overgrazing allows it to establish itself extensively excluding heather or grasses.

Braird. The first sprouting shoots of corn or other crop. Also called breer or brere.

Brahman. Indian or Zebu cattle characterised by loose skin on the throat and dewlap, with well developed seat pores, a muscular hump over the neck and shoulders, and large drooping ears. Commonly steel-grey in colour, but variations from black to white are found, and some red strains. They are long-lived cattle, tolernt of high temperatures, and resistant to insects and various tropical diseases.

Brake. 1. A harrow. 2. A frameworkd in which a restless horse is confined during shoeing. 3. A light carriage for breaking a horse to harness. 4. Brushwood or thicket. 5. An implement for crushing or braking flax or hemp. Also cattle a flax brake. 6. A scissor-like tool for stripping willow bark for basket-making.

Brake Harrow. A heavy harrow for breaking up clods.

Braky. Full of braken or brake, or rough and thorny.

Bramley. A well known variety of apple.

Bran. Thus husks of ground corn separated from the flour by bolting. A palatable feedingstuff with a useful protein and fibre content, especially for cattle and poultry. It is often fed to sows prior to farrowing and, as bran mash, is given to sick animals with no appetite. Also used in making brown bread.

Brand. 1. A mark burned onto an animal's hide for identification purposes by means of a hot branding iron. 2. General term for Blights or fungal diseases of grain crops.

Branding Equipment. An equipment which is used for marking of cattle, horses and sheep.

Brash. 1. Loose disintegrated rock and rubble, or a soil containing many stone chippings or rock fragments. 2. Pruned branches of conifers.

Brashing. The pruning of the lower branches of young conifers between 15 and 20 years old up to a height of about 5 fit to allow better access into a plantation.

Brassica. The generic name for cabbage, and its related plants including cauliflower, broccoli, kale, savoy, Brussels sprouts, and for turnip and swede.

Bratting. A now obsolete method for protecting sheep during severe weather by use of a 'brat' or cloth tied round the body.

Bray. Raucous outcry of the donkey and mule.

Bray's Nutrient Mobility Concept. According to this, as the mobility of a nutrient in the soil diseases, the amount of that nutrient required in the soil to produce a maximum yield (the soil nutrient requirement) increases from a variable net value. It is determined by the magnitude of the yield and the optimum percentage composition of the crop, to an amount whose value tend to be a constant.

According to Bray :

$$\log (A\text{-}Y)=\log A\text{-}C_1 b\text{-}CX$$

Where Y represents the yield obtained when X units of a plant nutrient are added to a soil having b units of the same immobile but available from a nutrient.

A represents the maximum yield obtained under the given conditions.

C_1 represents a constant representing the efficiency of b for yields.

C represents the efficiency factor for X which is the added fertilizer form of the nutrient b.

Brawn. 1. A Boar 2. A meat preparation from cut, boiled and pickled pig's head and ox-feet.

Brawner. A male pig castrated after serving the sows. Also called a stage.

Braxy. A bacterial disease of sheep causing severe inflammation of the stomach, dizziness, exhaustion and loss of appetite. Sometimes induced by indigestion caused by eating frosted herbage.

Braxy Mutton. Meat from a sheep that has died to braxy. Also a general term for meat from a sheep having died of disease or accident.

Braxy Pasture. Grassland infected with *Clostridium septique.* (Braxy).

Breadcorn. Corn, from the flour of which bread is made.

Break. 1. To tame or train a horse to wear a saddle or for draught work. 2. A change of crop in a rotation programme. Thus an arable break could be corn grown for several years in a field following usually a root crop and to be succeeded by perhaps a ley. 3. To cut up an animal's body.

Breakaway. A stampeding animal.

Breaking Ground. Ploughing uncultivated or fallow land.

Breast. The Mouldboard of a plough.

Breast Plough. An ancient but simple plough, spade-like, with a long shaft and cross bar which was pressed against the chest. Used for paring turf.

Breeching. A strong leather strap on a harness passed round a horse's haunches, attached to the saddle and the shafts of a cart, allowing the horse to reverse the cart.

Breed. 1. To reproduce (both animals and plants). 2. To promote the reproduction of animals and plants, often under control, in order to select certain characteristics for transmission to offspring. (Bloodline, Cross-Breeding, Plant Breeding). 3. A strain, race, variety, stock or kind of plant or animal. Mostly the result of a continuous cycle of hybridisation followed by inbreeding of an isolated pocket of plants or animals.

Breed Count. Microscopic method of counting bacteria in a dried, stained film of milk.

Breed in, Breed out. To introduce or remove a characteristic from an animal breed or plant variety by continuous breeding of those individuals lacking or having the characteristic, respectively, until it becomes fixed in or is lost from the breed or variety. (In Breeding, Character).

Breeding Crate. An apparatus designed to take the weight of bull which is too heavy for the heifer or cow with which is to be mated. Also called a service pen or service crate.

Breeding Stock. Farm animals selected for producing offspring, as opposed to being fattened for slaughter, etc., in order to maintain or increase the size and quality of a herd of flock.

Brewers' Grains. The residue of barley after being used in brewing beer, consisting mainly of protein and fibre, used in either wet or dried as an animal feedingstuff. Also called Draff.

Bridge Grafting. Refers to a form of repair grafting for plants whose root system is damaged but bark of the trunk is injured. Trees of some species, like the cherry, and pecan can heal over extensively injured areas by the development of callus tissue.

Bridle. An apparatus fitted to a horse's head to control it, including the bit and reins.

Brim. 1. A sow on heat. 2. To put a boar to a sow for mating.

Bring Off. To hatch out eggs.

Brisket. The breast of an animal. The term is used for the cut of meat from next to the ribs.

Birtch. The thigh and twist part of a sheep, and also the wool from it which is coarse and of low quality.

Brittle Hoofs. Refer to an abnormally dry state of the horn. The hoofs tend to become almost of the consistency of stone, chip and crack easily, cause contracted heels, and lead to difficulties in shoeing. Long-continued dryness or stabling on dry, hard floors has been conductive to the trouble.

Brix. Refers to the percentage of total solids in sugarcane juice. It is read off from brixometer ; range from 18 to 22 ; 15 in juice of young and immature cane.

Brix Saccharometer. An instrument which is used for measuring the degree and gives the direct reading of the total solids dissolved in the juice. It has been found to vary from 22 to 225 depending on the variety.

Brixometer. Special hydrometer which is calibrated to read off directly the quantity of total solids in sugarcane juice.

Broadcast. Means sow or scatter seed on the surface of the land by hand or by machinery.

Broadcast Fertilizer Distributor. A machine which broadcasts fertilizer. Various distributive mechanisms exist including spinning discs, revolving fingers, rollers, brushes and chains etc.

Broadcaster. A machine for sowing seeds, mainly grass and clover, consisting usually of a hopper supplying seed to one or more revolving brushes by which it is scattered (Fiddle).

Broadleaf. 1. A term applied to non-coniferous trees, almost coterminous with deciduous. Leaves are usually broad, flat and then, as opposed to linear needles, and veins are networked as opposed to parallel. Broadleaf trees or hardwoods which are slower growing than Conifers and produce compact hard wood, take longer to provide a profit when grown commercially. They are suited best to lower hill slopes and deep fertile lowland soils. 2. A term applied to dicotyledonous plants, usually weeds, e.g., dock, Fot Hen, e c.

Brocket. 1. A stag in its second year with its first horns, unbranched and dagger-like. 2. Mottled.

Broiler. Immature chicken of 8 to 12 weeks. Usually young male weighing from 3/4 to 2-1/2 lb, sufficiently tender to be broiled.

Broken Furrow. A type of furrow slice turned by a Digger plough, rough and broken, as opposed to the smooth continuous furow left by a Lea Plough.

Broken Horse. A horse trained to the saddle or Bridle.

Broken-mouthed. A descriptive term of an old sheep which is unable to deal with it food requirement due to loss of some teeth.

Broken-winded. Having short breath or defective respiratory organs and thus incapable of hard wark. Particularly applied to horses.

Broken Work. Ploughing which leaves the furrows broken. Also called broken rib work. (Broken Furrow, Whole work).

Brome Grass. A large genus (*Bromus*) of long-awned grasses strongly resembling oats, but more flowery, mostly unimportant as fodder and generally regarded as weeds in Britain and becoming increasingly troublesome.

Bronchitis. Refers to the inflammation of the mucous membrane of the bronchial tubes.

Bronchopneumonia. Refers to the inflammation of the bronchi and of the lungs.

Brood. 1. A family of birds hatched together. 2. To sit (as a hen) on eggs to hatch them. 3. The eggs, larvae and nymphs of bees in a brood chamber (Beehive).

Brood Cell. Cell in which the larvae of bees are reared.

Brood Chamber. Part of the beehive in which the brood is reared.

Brood Comb. A wooden frame enclosing a wax sheet from which bees build the cells of the Honeycomb, placed in the brood chamber of a hive. (Beehive)

Brood Frame. Wooden frame having the brood comb.

Brood Mare. A female horse kept for breeding.

Brooder. A unit containing a heat source in which newly hatched chicks are kept under controlled, gradually reducing temperature conditions throughout the Brooding Stage.

Broodiness. Pertaining to hen which attempts constantly to sit on eggs.

Brooding. Refers to the rearing of chicks after hatching till maturity; natural and artificial methods followed.

Brooding Coop. Coop or cage which has been provided to hen and the chicken to protect them from enemies and increment weather, essential requirements of which being drynets, roominess and safety; bamboo coops, barrels, packing boxes and A-shaped coops commonly used.

Brooding Stage. The period, 4-8 weeks, from hatching until poultry chicks cease to need some form of heat for survival, provided by a Broody Hen under the protection of its wings, or artificially in a Brooder.

Broody Hen. A hen inclined to sit on its eggs to hatch them. Also called a sitter.

Brow Antler, Brow Tine. The first tine of a deer's horn.

Brown Dog-Tick. A pest of dogs which is sometimes found on livestock.

Brown Forest Soils. An intrazonal group of soils having very dark brown surface horizons which have relatively rich humus (mull) grading through lighter colour soil into the parent material and characterised by a slightly acidic reaction, little or no illuviation of iron and alumina, and moderately high content of calcium in the soil colloids. Such soils have been developed under deciduous forest in temperate humid regions from parent material relatively rich in bases.

Brown Heart. A Deficiency Disease of swedes due to lack of boron, causing browning or mottling of the root making it unpalatable.

Brown Oil of Vitriol (B.O.V.). Sulphuric acid, formerly used as a weed-killer, particularly against broadleaves such as charlock.

Brown Rot. A fungal disease of apples, pears and plums, which turns the ripe fruit brown and rotten. They become mummified, hanging on the trees.

Browse. 1. To feed and nibble on young shoots, buds and twigs of plants. 2. The tender shoots of shrubs and trees which cattle may feed on.

Bruccila Abortus Vaccine. It is obtained from a strain of *Brucella abortus* whose virulence has been reduced. It is widely used for the vaccination of calves between 4 and 8 months of age and is highly effective in preventing brucellosis.

Brucellosis. A bacterial disease of the reproductive system, mainly of cattle sheep and pigs, although all mammals have been susceptible. In cattle it causes infection of the womb and placenta and abortion of calves. It is spread mainly via aborted calves, afterbirth, discharges and milk which contaminate pasture, food and water. It causes undulant fever in humans. Also called contagious abortion.

Brucellosis in Pigs. A bacterial disease which is charaterized by abortion, still births, abscesses in different parts of the body and temporary or permanent infertility in both males and females. It occurs due to *Brucella suis*, a short slender 1-2_ long and 0.3-0.5_ wide gram-negative, non-capsulated, non-motile and non-sporulating organism.

Brush. Loppings and trimmings from trees, shrubs and hedges. Also to lop or trim such plants, or to cut down weeds using a Brushing Hook.

Brushing Hook. A sharp sickle-shaped tool which is used to trim hedges and clear undergrowth.

Brushwood. Low, rough, scrubby, undergrowth or thicket.

Brutting. A method of summer pruning used mainly by nutgrowers, but also applied to old fruit trees, in which side shoots of the bushes are snapped off near to their base by hand, and the ends left hanging, preventing secondary growth.

Bubonic Plague. Plague which is charaterized by buboes (swelling of glands in the groin). It primarily affects, rondents, but is often transmitted to man and livestock.

Bucket Elevator. An equipment which is used to lift the seeds to the top of bins/machines for cleaning, sorting etc.

Bucket-feeding. Feeding a young animal with milk and gruel from a bucket, rather than allowing it to suckle its mother.

Buckhound. A small staghound used for hunting bucks.

Buckling. A male goat, 1-2 years old.

Buckrake. A simple implement with close-set long tines, usually rearmounted on a tractor hydraulic lift linkage, used for collecting and transporting cut grass and Green Crops for Silage making. Also used for other purposes such as transporting straw and hay bales.

Buckshot. A large type of shot used in dear-shooting.

Buckwheat. Two frost-sensitive cultibated plant species, common buckwheat (*Fagopyrum esculentum*) and Tartarian buckwheat (*F. tartaricum*), often classed as Cereals, but actually of the dock family. Grown mainly in patches, providing grain for peasants, and sometimes as a forage crop. Common buckwheat has white or pink flowers and dark brown seeds, sharply triangular in cross-section. Also called brank.

Bud Mutation. Mutation in the bud which brings about variation in it and produces a branch, flower or fruit unlike rest of the plant.

Bud Union. The term used for the place of union of the bud with the plant-stock in a budded plant.

Budding and Grafting Knives. Knives which find use in budding and grafting operation in the asexual methods of plant propagation.

Budsports. Branch showing changes in one or more inheritable characters that could be perpetuated by asexual means.

Budstick. Shoot whose buds have been removed for preparation.

Budwood. Plant material which is used in budding of fruit and other trees. It is obtained from healthy vigorous and high yielding trees or branches which generally bear fruit of good quality, taken from mature wood of current year's growth, and from only round twigs, especially in the case of citrus species, the best being those from basal portions of young shoots.

Buffalo Grat. Small insect which is belonging to the Simulium genus. It is often a terrible scourage to cattle and horses. It could be controlled by spraying D.D.T.

Buffer Soil Compounds. Clay, organic matter, and such compounds as carbonates and phosphates which make the soil to resist appreciable change in pH value.

Bug. Term strictly used only for members of the order Homoptera.

Bug Spittle. Nymph of a sucking insect of the family Cercopidae. It commonly lives surrounded by a white, frothy mass resembling spittle.

Bulb Dips. Chemical compounds in which bulbs are immersed to provide protection against pests and diseases.

Bulb Syrinage. It consists of a soft rubber bulb with a hard rubber pipe to be introduced into body openings *e.g.*, into the vagina.

Bulbar Paralysis. Refers to the paralysis of the prolongation of the spinal cord into the brain.

Bulk. The dry matter in animal rations.

Bulk Bin. Large container which is used in handling and strong fresh fruits and vegetables. Most bulk bins have a capacity of 10 to 20 bushels.

Bulk Density. Refers to the mass or weight of oven dry soil per unit bulk volume, including air space. This mass in relation to the weight of a unit volume of water was formerly termed as Apparent Density or Volume Weight.

Bulk Method. Method which is used in self-polinated crops in which the segregating generations of hybrids have been grown in bulk plot with or without mass selection, and with single plant selection in F_6 and later generations. Synonyms, bulk breeding and bulk method of breeding.

Bulk Seed Sample. Entire quantity of sample which is received in a seed laboratory for testing.

Bulk Specific Gravity. Refers to the ratio of the bulk density of a soil to the mass of unit volume of water.

Bulky Feedingstuff. A feedingstuff having little nutritive value relative to its volume, e.g., hay, straw or chaff in contrast to concentrates.

Bull. An uncastrated aduly male bovine animal.

Bull Beef. Beef from a bull as opposed to a steer. A bull carcase provides more lean meat than a steer's.

Bulldog. 1. A deformed calf with a bulldog-like appearance due to its short legs and jaws and swollen abdomen. 2. A split ring which grips a bull's nostrils through which a rope can be passed for holding or leading.

Bull Dog Calf. Calves born dead, usually prematurely; and the anatomical alterations, including short legs and bulldog-like head are common.

Bulla. Large sac like structure which is filled with fluid under the skin.

Bullet. A dose of mineral, e.g., cobalt or magnesium, introduced to an animal by a special 'gun', providing it with a long-lasting supply, preventing Deficiency Diseases.

Bullimong. A mixture of oats, peas and mixed grains shown as a Forage Crop. The custom was confined to certain limited areas and has probably now ceased.

Bulling Heifer. A maiden Heifer which has reached the right size and age for mating.

Bullock. A castrated bull. A bullock calf is a castrated male calf.

Bull-wheel. A large wheel which is providing drive to the machinery of a binder or reaper.

Bumble Foot. A condition of poultry characterised by abscesses between the toes, caused by thorns, stones, glass fragments, etc., penetrating the soft tissue, or by bruising, and resulting in lameness.

Bunch. 1. A handful of Hay. 2. A herd.

Bund. Low ridge of soil which is built along the field contours.

Bund Former. An implement which is used for making bunds or ridges by collecting the soil. Bunds are needed to hold water in the soil, thereby conserve moisture and prevent run-off.

Bunostomum. Hookworms found in cattle, sheep and goat.

Bunt. A fungus disease (*Tilletia caries*), mainly of wheat but also of rye, filling the ears with a mass of black, fish-smelling, greasy spores. Bunted grains burst when threshed, the spores contaminating and discolouring healthy grain, so that it is useless as seed or for milling. Also called stinking smut.

Bur, Burr. 1. The prickly seed-case or fruiting head of certain plants, particularly burdock, which adhere to animal fur and clothing. 2. A knob at the base of a deer's horn. 3. A knotting growth on a tree, leaving marks in timber. 4. The catkin or cone of the hop.

5. Arenaceous rock from which millistones (burr-stones) are made. Also one of the corrugations of a millstone.

Burgundy Mixture. A misture of copper sulphate and sodium carbonate (washing soda) used as a fungicide. (See Bordeaux Mixture).

Burnt Lime. A form of Lime with the highest neutralising value. Mainly Calcium Oxide and various residues produced by burning lumps of Chalk or Limestone, and usually kibbed or ground before use. It readily absorbs water to form Calcium Hydroxide and cannot be stored for very long. Caustic and irritant, it is also used to destroy carcases of animals which have died from infectious diseases. Also called quicklime, lump lime and shell lime.

Burry Wool. Fleeces which have caught seed Burs in them.

Bush. Term which is used to signify a region covered with forest or scrub not yet cleared for cultivation.

Bush-and-Bog. Refers to heavy cut-away disktillage implement used on rough or brushy pasture land.

Bush Disease. Sporadic Bovine Encephalomyclitis.

Bush Drain. A type of field Drain which is constructed by packing bushes into a trench which is then refilled.

Bush Fruit. Fruit which is grown on bushes as opposed to trees or Cane, e.g. blackcurrants, gooseberries.

Bush Fruit Trees. A fruit tree, the lowest branches of which are below 75 cm (30 in) above the ground.

Bush Harrow. An old type of light harrow having a barried wooden frame with bushes or branches woven through it, used for covering grass seeds. Developed from the primitive harrow with a log tied on top of a bush.

Bushel. A dry measure by which grain was once generally computed, and also used for fruit, containing 1.28 cubic feet or 8 gallons. Many farmers still Drill seed in bushels per acre or refer to grain in terms of lbs per bushel when discussing its quality, but, following metrication, grain traders now use kg per hectolitre. Fruit is now usually picked into bulk bins rather than bushel boxes. Also called sieve.

Butcher System. A design, once popular, now rare, for hop garden wire-work, with three horizontal wires attaching to the poles the lowest at about 15 cm (6 in.) and the middle (breast wire) at about 120 cm (4 ft.).

Butt. 1. A tree trunk. 2. The base of a leaf stalk. 3. One of the short furrows ploughed where a field runs to an angle or at an awkward corner.

Butterfat. The fatty substances contained in milk, mainly glycerides of palmitic and oleic acids, present as minute fat globules, the size of which vary according to breed, and diminish from time of calving. The largest and present in Jersey milk, churn easily and are best for butter-making. The smallest are present in Ayrshire and Freisian milk and are best for cheese. The butterfat content of milk is determined by the Gerber Test and nowadays also by a number of semi-automatic instruments (e.g. Foss Milkotester) which are themselves calibrated using the Gerber Test.

Butterfly. A breed of rabbit.

Buttermilk. The residual liquid following the churning of cream into butter.

Byre. A cow-house.

C

C. The ratio expressing relative tail length of the nematode. It is calculated by dividing the total length by length of the tail.

C-4 Pathway. A cycle reported in a few tropical and temperate plants. It is characterized by using higher light intensities for fixing CO_2 than does the Calvin cycle.

CD : (Critical Difference). *See* LSD.

C : N Ratio. Carbon-nitrogen Ratio.

C-Horizon. Soil horizon.

C Trials. Series of microplot trials which are carried out for determining the value of a new variety over a wide area of possible usefulness.

C_3 Plant. A plant in which the first product of carbon dioxide fixation has been the 3-carbon compound phosphoglyceric acid. Examples of this group include most crop plants except the grasses.

C_4 Plant. A plant in which the first product of CO_2 fixation has been the 4-carbon compound oxaloacetic acid. Examples of C_4 plants include many tropical grasses, corn, sugarcane and some weed species.

Caatinga Forests. Forests which are found in Brazil. In these forests the leaves fall in the dry season.

Cabbage. Abiennial cruciferous plant which is grown both as Fodder and for the table. It is characterised by a short steam with a very large head comprising tightly overlapping leaves. Cabbages are commonly fed to cattle, usually after milking, and pigs.

Cabbage Lettuce. A type of lettuce which resembles a cabbage both in shape and its possession of a distinct heat.

Cabbage Root Fly. A fly (*Erioischia brassicae*), which resembles the housefly. It lays white eggs in the soil surface near the stems of seeding Brasricas. The maggots, which hatch within a week, feed on the young roots and stems, causing plant water loss and frequently death.

Cabbage White Butterfly. A common butterfly (*Pier is brassicae*) having white wings, and yellowish-green, black-spotted caterpillars which feed on cabbages and other Brassicas. Also known as Large White Butterfly.

Cacti. Nearly all its species may be grown in a moderately-heated green house.

Cade. A lamb or colt reared on the bottle. Also called Cosset Lamb.

Caecum. Refers to the blind, sac-like end of the large intestine which is connected to the small intestine. Also called blind gut.

Cage Wheel. Wheel or an attachment to a wheel having spaced cross bars to improve the reaction of the tractor in a wet field. It has been generally used in paddy fields.

Cake. Refers to the seeds of various plants, mainly tropical (*e.g.,*groundnut, cotton, soya bean, coconut, linseed and palm nut kernel) which is compressed (to extract oil for cooking, soap manufacture, etc.) into flat slabs, or straights, rich in Protein, having oil and fibre in verying amounts, used as cattle food.

Calamintha Alpina. A hardy aromatic plant, lorming a thick carpet of coliage and carrying sprays of violet flowers all through the summer.

Calcareous Soil. An alkaline soil which is having sufficient calcium and magnesium carbonate to bring about visible effervescence when treated with hydrochloric acid.

Calcification. Refers to the process of the deposition of insoluble lime salts, especially calcium carbonate or phosphate in tissues (2) Refers to the process of accumulation of a layer of lime in the soil profiles in a dry climate, as it is happening in the soils of Rajasthan.

Calcified Nodules. The nodules which are caused by nodular disease. They occur in knotty guts.

Calcium. (Ca) A soft white chemical element which is essential to life, and considered to be an important constituent of bones and teeth, and of the several compound forms of lime used in agriculture (Calcium Carbonate, Calcium Hydroxide and Calcium Oxide).

Calcium Carbonate. ($CaCO_3$). A white insoluble solid which is found naturally as chalk, limestone, marble and calcite.

Calcium Cyanide. A very effective fumigant to control, insects and rodents.

Calcium Cyanamide. ($CaCN_2$). Nitrolime. An artificial fertilizer which is powdery or granular, blue-black in colour, supplying lime and nitrogen to crops. It is converted by soil water to ammonia.

Calcium Hydroxide. $Ca(OH)_2$). Slaked lime or hydrated lime. It is in a finely divided condition by horticulturists to correct soil acidity.

Calcium Levulinate. White powder which is soluble in water and therefore used increasingly for the preparation of sterile solution.

Calcium Oxide. (CaO). Quicklime. A lumpy white powder. It is caustic and irritant and is used to destroy carcases of animals which have died from infectious diseases. Also called burnt lime, and lump lime.

Calcium-Phosphorus Ratio. Normal ratio of calcium to phosphorus in feed ratios has been between 2:1 and 1:1. The ratio 2:1 is generally preferred. Vegetable protein concentrates and poor in calcium and phosphorus.

Calcium Sulphate. Gypsum.

Calender of Operations. Refers to graphic or tabular presentation which shows the kind of farm operations to be carried out during the season and specified time limits within which work has to be carried out.

Calendula. Hardy annuals which thrive in beds or borders and is almost any soil.

Calf. The offspring (in its first year) of a cow. A male is known as a bull calf, a female is known as a heifer calf, quey clf or cow calf.

Calf Diptheria. Refer to an acute, infectious disease which is characterized by the formation of a diphtheric, false membrane on the mucous lining of the mouth and throat of young suckling calves. The mortality from the disease has been very high. It has been caused by *Actinomyces necrophorus*.

Calf Knee. Term is used for bone abnormality which has been observed especially in horses.

Calf Pneumonia. Affecting calves ranging from 15 to 75 days of age. It arises suddenly. The affected animal manifests dry cough, extreme lacrimation and a temperature range of 104.8° to 105.5°F.

Calf Scours. An important bacterial disease (often *E. coil*) of both dairy and beef calves, often fatal. Also known as neonatal diarrhoea.

Calf Scours Serum. It prevents the heavy losses from white scours if given to calves soon after their birth.

Caliche. More or less cemented desposit of calcium carbonate often mixed with magnesium carbonate at various depths. It is the characteristic of many of the semiarid and arid soils of the Southwest.

Calk, Calkin. A pointed, turned-down piece on a horse-shoe which prevents slipping.

Callus. 1. Refers to the protective tissue which grows, as a swelling, over wounds in woody plants. 2. The cartilaginous growth, having collagen, which first grows to heal a bone fracture, later becoming bone tissue. 3. Any thickned tissue which is often hardened or leathery.

Caloric Energy of Feed. May be defined as the inherent power of feeds for supplying nutrients which are essential to the reproduction, maintenance, growth and milk production of animals.

Calorie. The quantity of heat required to raise the temperature of 1g of water by 1°C. A kilo-calorie (K.cal) or Large Calorie (Calorei) is 1000 calories.

Calve, Calve Down. To give birth to a calf.

Camber. An Auto-Sex-Linked breed to poultry, with barred feathers, a single comb, yellow legs, and producing white eggs.

Cambium. A layer of actively dividing cells in plants which produces xylem, or wood, to its inside and phloem, or bast, to its outside in the process of secondary thickening. It is located just beneath the bark of trees and shrubs.

Camshaft. Shaft in 1 C engine which is able to raise and lower the inlet and exhaust valve at proper time. It is mainly used to operate the ignition timing mechanism, lubricating, oil pump and fuel pump.

Canal. Artifical watercourse which is usually cut.

Canal System. A system which comprises of all the canals. It constitutes a complete irrigation system.

Cancer. Malignant tumour.

Cancer Eye. Malignant tumour which primarily attacks the eye of cattle and may be caused by irritation of the eyes by dust, sand, insects or strong rays of the sun. Its treatment involves the total removal of the cancerous tissue togheter with the eyeball by an experienced veterinarian.

Candied Fruit. Fruit impregnate with sugar, drained and dried is known as candied fruit. It is not sticky and is plump, tender and exceedingly sweet with high flavour.

Candling. Refers to a process by which eggs are allowed to pass over a candling machine in which a light source detects blood spots and shell cracks. The apparatus is usually combined with an egg grading machine.

Cane. The stem of reaspberry blackberry and other similar plants, and of the larger grasses, such as sugarcane or bamboo.

Cane Sugar. A type of storage sugar (Sucrose) present in sugar cane, sugar beet, ripe fruits and tree sap (*e.g.*, maple sugar). It is broken down by intestinal enzymes into glucose and fructose which can then be absorbed by the bloodstream to provide energy.

Canine. Family of animals including dogs, cogotes, foxes and jackals etc.

Canker. 1. A general term which is used for a number of fungal diseases of tree (often due to *Nectria sp.*), especially those affecting fruit trees, in which growing tissues are destroyed and cortical tissues malformed, forming cancerous growths and open wounds. 2. A fungs disease which causes softening of the horn of a horse's hoof, inflammation and the production of a soft chessy material. 3. A colloquial term for ear mange in dogs and for the pussy conditions which can develop if not treated.

Cannibalism. The term used for the practice of an animal eating its own kind. It sometimes takes place, with indoor-housed poultry due to overcrowding, boredom (scratching for insects is precluded in cages) our protein deficiency, and has been controlled by removing the point of the upper beak (debeaking) or the use of soft red lighting. Sows will also occasionally eat their own or other sows' piglets under intensive breeding conditions.

Canning. Refers to the process of preserving food by the application of heat high enough for destroying essentially all microorganisms present together with sealing the food in air-tight sterilized canes so as to disallow recontamination and to preserve the food as early as possible in the condition in which it could be served when freshly cooked. Fruits are canned in tin cans or glass jars.

Cannon Bone. A bone supporting the limbs of horses, between the knee and fetlock of the forelegs, and between the hock and fetlock of the hind legs.

Cannula. A tube having a sharp-pointed plunger (a Tractor) which has been inserted to puncture an animal's rumen. The cannula is left in position when the trocar has been withdrawn to allow gas to escape.

Canopy. Refers to branches, leave etc. which are formed by woody plants at some distance above the ground.

Canopy Class. Usually refers to a classification in accordance to completeness of canopy, *see* **Canopy Density.**

Canopy Density. Refers to the relative completeness of the canopy. It is usually expressed as a decimal coefficient, taking closed canopy as unity. The following classification of canopy density is in vogue : *closed* if the density is 1.0 : *dense* if the density is between 0.75 and 1.0 ; *thin* if the density is between 0.5 and 0.75 and *open* if the density is less than 0.5.

Cant. 1. A sloping bank. 2. An area of woodland clear-felled at one time. Often the unit of sale (by auction) for sweet chestnut *Castanea sativa*), commonly about an acre in extent. 3. An area which is designed to a ploughman to plough in a ploughing competition.

Canterbury Hoe. A type of hoe having three long, flat, sharp progns. Used break up clods.

Capacity, Field Moisture. Amount of water which is expressed in percentage of dry weight, that a soil is able to retain against the pull of gravity *i.e.* after saturation and allowing excess water to drain away.

Capabillty Class, Soil or Land. A rating which indicates the general suitability of a soil or land for agricultural use.

Capillary Action. Refers to the phenomenon of a liquid rising up a narrow tube, or the formation of drops, bubbles and films, due to high inter-molecular attraction within the liquid. Important in soil as the natural forces of capillary attraction cause water to rise through the minute 'capillary space' between soil particles and through soil pores.

Capillary Fringe. Refers to the water in the layer of soil or subsoil into which ground-water is entering due to capillary rise. The effective height of the capillary fringe is dependent on the pore sizes of the soil.

Capillary-Fring Belt. Refers to that portion of the aeration zone which is situated between the intermediate belt and water table.

Capillary Migration. Refers to the movement of water through a rock or soil material for the water. It takes place in the zone of aeration.

Capillary Potential. The term used for the measure of the attractive forces with which water is held by soil.

Capillary Water. Refers to the water which is held in the soil in pores, and capillary spaces, and as a thin film surrounding soil particles, held under tension against the pull of gravity. Capillary water moves from water to drier parts of the soil under hydrostatic gradient. It is actually the 'permanent' soil solution having dissolved substances such as applied plant nutrients.

Capillary Zone. Zone in which soil water is held by capillary forces. The interstices may be completely filled (Saturation zone) or partly filled (aeration zone). Pressures within this zone have been less than atmospheric

Capon. A castrated cockeral. Nowadays caponisation is carried out chemically by injecting female hormones (oestrogens) which make the testes to regress. It also produces a more tender carcase and stops cockerels crowing and fighting.

Capped Blossom. Apple blossom in which the petals fail to open and from a brown 'cap' over the buds, as a result of the Apple Blossom Weevil eating the heart of the blossom.

Cappie. A disease of sheep (mainly old lambs and young sheep), generally those grazing hill areas in autumn and winter, considered to be due to phosphorus deficiency. It is characterised by lack of thrift and thinning of the skull bones. In bad cases sheep are unable to eat or close the mouth. Also known as double scalp.

Capping. The term used for the formation of a crust on a soil during a hot, dry spell, usually after heavy rain which makes the surface soil particles to slake down, blocking up the soil pores.

Capped Hock. Refers to the firm swelling which appears on the point of the hock of livestock. This blemish may be of the size of an apple or so small that it escapes notic. It occurs because of constant irritation, such as may be produced by rubbing or kicking the walls of the stable. The condition is not considered serious because it is not accompanied by lameness.

Caprifig. Refers to the race of fig that does not form edible fruit but gives food for the wasps which pollinate the figs.

Capsid. Protein coat of a virus particle, which consists of capsomeres.

Capsid Bug. Refers to one of a large family of greenish or brownish plant Bugs, slender and elongate, about 6 mm (¼ in.) long. The common Green Capsid is a pest of fruit and vegetables. The Apple Capsid which preys on the stem tips, fruit and foliage of apple trees may be now largely controlled by D.D.T. and other insecticides.

Capsule. A type of hard, dry, dehiscent fruit which is opening in various ways to release the seeds, *e.g.*, poppy. 2. A gelatinous envelope which is founder around certain kinds of bacteria. 3. A soluble gelatinous case having a dose of medicine, administered orally to an animal. 4. The fibrous membrane which envalopes the various organs of the body, *e.g.* liver, spleen. 5. A thin metal envelope in an incubator, having a liquid (*e.g.*, ether) which expands and contracts, maintaining an even temperature regime in the incubator.

Captan. A fungicide which is used in seed dressings for peas and vegetables. Also used effectively against apple scab.

Carbamates. A group of alkaloid insecticides which are selective and to some extent systemic in their action, and are non- persistent in the environment, *e.g.*, carbaryl, methiocarb.

Carbohydrates. A large group of compounds having carbon, hydrogen and oxygen only, of the general formula $Cx(H_2O)y$. They are produced by plant Photosynthesis and are essential to metabolism in all living organisms.

Carbohydrates form the largest part of the food of animals which have only small amounts of glycogen and sugars in their bodies. The complex polysaccharides in plants are broken down by animals into simpler sugars by digestion before they can be absorbed and used. The energy stored in carbohydrates is released, to power living processes and provide heat, by 'burning' with oxygen during cellular respiration.

Carbolineum. Wood preservative which is prepared from coaltar It consists of anthracene oil and is an effective disinfectant and insecticide. It is widely used for the preventation or control of mites and ticks infecting stables, poultry houses, barns etc.

Carbon. (C). A chemical element. The principal constituent of all organic compounds and essential to life.

Carbon Assimilation. Photosynthesis.

Carbon Cycle. The circulation of carbon atoms between living organisms and the atmosphere. Carbon dioxide is assimilated during Photosynthesis by plants to produce carbohydrates and other complex carbon compounds. Animals feed directly on plants or on herbivores to obtain these compounds which are broken down during both animal and plant respiration to release carbon dioxide back to the atmosphere. Carbon dioxide is also released from the decaying remains of plants and animals by the action of bacteria and fungi.

Carbon Dioxide. (CO_2). A colourless, odourless, gas occurring in the atmosphere. It is utilised by plants in Photosynthesis to produce Carbohydrates. A by-product of animal and plant Respiration and combustion (*e.g.,* straw burning).

Carbon-Nitrogen Ratio. Refers to the ratio of the weight of total organic carbon to the weight of total nitrogen in a soil or in an organic material. The C:N ratio of wheat straw has been nearly 70:1 where as that of soil is 10:1. When undecomposed straw with a high C:N ratio is applied to the soil, its C:N ratio gets reduced through bacterial decomposition. To speed up bacterial decomposition, nitrogenous fertilizer has been added to the soil at the time of turning under straw.

Carboxydismutase. Enzyme which is responsible for the fixation of inorganic CO_2 into organic compounds in the dark reactions of the C_3 photosynthesis cycle; also known as ribulose diphosphate carboxylase.

Carbon Tetrachloride. (CCl_4). A sweet smelling, colourless liquid which acts against liver flukes and round worms in animals (excluding cattle for which it is not sale). It is usually administered orally in gelatin capsules.

Carcase, Carcass. A dead body. In butchers' terms it refers to an animal's body after removal of the head, limbs, hide and offal. Also known as a dressed carcase.

Carcass Residue (Mammals). Residue which is obtained from animals tissues exclusive of hair, hoofs horns and contents of the digestive tract.

Cardinal Temperatures. Minimum, optimum and maximum temperatures which are affecting plant growth. The minimum temperature is defined as that point at which growth begins; the optimum temperature is that point at which growth is best; and the maximum temperature is defined as that point at which growth stops.

Carminative. Drug relieving flatulence and colic by stimulating the movements of stomach and intestine.

Car ivore. Any flesh-eating mammal which has teeth and claws *e.g.*, dogs, cats, bears etc.

Carnivorous Plants. Plants which are able to capture small insects or animals and feed upon the nitrogenous compounds derived from them *e.g.* pitcher plant, bladderwort etc.

Carol Bean An alkaloid-containing poisonous plant.

Carotenes. A group of reddish yellow pigments. They are unsaturated hydrocarbons ($c_{40}H_{56}$), occurring in carrots and butter, providing their distinctive colours. Carotenes are widely distributed in plants, acting as photosynthetic pigments (Photosynthesis) in cells lacking chlorophyll, and are converted into vitamin A by animals.

Carotenoid Pigments. Orange or yellow pigments which are occurring in cells are crystals, usually in chromoplasts or in plastids.

Carotenoids. A group name for lipochromes or carotenelike plant pigments (Yellow compounds), which through food, get deposited in animal tissues.

Carpus. The human wrist, or corresponding part of an animal's forelimb containing the animal's carpal bones, usually called the knee.

Carr. Boggy ground or land reclaimed from such areas by drainage (See Land Drainage). Also a fen wood, common in East Anglia, dominated by alder.

Carrier. An animal possessing or 'carrying' the germs of a disease, having itself recovered and showing no symptoms but, nevertheless able to pass the infection on to other animals.

Carry Over Soil Moisture. Moisture which is stored in crop root zone between cropping seasons or before the crop gets planted. This moisture contributes to the water needs of crops before they are irrigated.

Carrying Capacity (Wildlife). Refers to the maximum number of a given wildlife species which in any given territory will support through the most critical period of the year.

Carrot. A root crop bred from the wild carrot (*Daucus carota*) with a long, tapering, reddish or yellowish root, which is sweet and edible. Grown mainly for human consumption, often on contract, for canning or freezing. Surplus or unmarketable roots are used as stook feed, mainly for cattle.

Carrot Fly. A small pestilent fly (*Psila rosae*), green- black in colour, attacking carrots, cellery, parsley and parsnips. Eggs are laid in the soil near the plants and the slender white larvae burrow into the roots near the growing tips. Growth is retarded and foliage loses its freshness. With serious infestation roots may fork or show distorted growth, and in some cases crops may fail.

Carry-on-period. The stage in poultry rearing following brooding and before chicks are able to survive without assistance.

Casein. Milk protein. It forms yellowish white powder, soluble in diluted acids or alkalies.

Caseous. Cheese like.

Cash Crop. Crop grown for sale which brings cash money immediately e.g., tobacco, cotton, sugarcane etc.

Caste. Refers to one of the distinct forms found among certain social insects.

Casting (Ploughing). If a plough works round a strip of unploughed land it is termed to be casting.

Castor Oil. An oil extracted from castor bean. It is a pale yellow liquid and is widely used as a mild purgative. It is sometimes effective for the removal of small wart *e.g.,* those on udders of cows.

Castor Oil Mixture. It consists of varying quantities of castor oil and neutral oil. This castor oil mixture is administered in connection with the treatment for the removal of palisade worms, pin worms etc.

Castrate. Means removing the testicles or to destroying this use.

Castration. Refers to the process of surgical removal of the testicles of a male animal or cutting of the vas difference by applying pressure with specially designed forceps such as Burdizzo forceps. It is carried out when calf is 8 to 18 months old. Early castration makes animal to develop into longboned or rangy type of bullock, while late castration makes animal to develop a heavierboned, short and thick-set bullock.

Casual. Plant occurring in a plant community of which it is not a regular inhabitant.

Catapult Mechanism. Refers to the method of seed dispersal which depends on the jerking of a long stalk swaying in the wind.

Cataract. Refers to the loss of transparency of the eye's lens.

Catarrh. Refers to the inflammation of a mucous membrane especially of the air passages of head and throat or of the stomach.

Cath Crop. (1) Quick-growing crop which could be planted and harvested between two regular crops in consecutive seasons, or between patches of regular crops in the same season (2) Crop which is planted after another has failed or when the season has been too late for the usual crops.

Catechu. Name used for two drugs both of which are astringents obtained from trees. They have been compatible with zinc, calcium or iron compounds. Catechus are sometimes used in the treatment diarrhoea.

Catena. Group of soils always occurring in association with one another in some characteristic pattern related to topographical factors.

Caterpillar. The larva of a butterfly (*e.g.* Cabbage White) or moth, often causing damage to crops by feeding on foliage.

Catgut. It is produced from sheep's intestines and finds use in surgery for tying blood vessels or stiching tissues together. Catgut gets absorbed by the system. It must be kept and handled under a septic conditions.

Cathartic. A medicine acting by stimulating the movements of the intestines or by liquifying the intestinal contents. Mild cathartics like sulphur, mineral oil etc. are called laxatives while more powerful cathartics like castor oil, epsom salt and also are purgatives.

Cation. A positively charged Ion.

Cation Exchange. The interchange of a cation in solution and another cation held on the surface of clay, and organic, colloids in the soil. For example, in acid soils subject to leaching, calcium ions have largely been displaced and replaced by hydrogen ions. The addition of lime corrects the acidity by exchanging calcium ions for hydrogen ions. Temporary sea water inundation of farmland replaces soil calcium ions with sodium ions. Applications of gypsum (calcium sulphate) corrects the alkalinity by exchanging calcium ions for the sodium ions. Cations applied in the form of fertilizers are absorbed by soil colloids, usually being exchanged for hydrogen ions. Nutrient ions in the soil solution taken up by plant roots are replaced by exchange able cations from the soil colloids.

Cation Exchange Capacity (C.E.C.) The total amount of exchangeable cations that a soil can absorb, expressed in milliequivalents per 100 g of soil or of other adsorbing material such as clay. It is a measure of the potential of a soil to hold nutrient cations for plant absorption. Clayey or organic soils usually have a large C.E.C., whilst the opposite applies to sandy or weakly organic soils. Agriculturally productive soils require application of fertilizers rich in cations and the C.E.C. is a guide to the quantity and frequency of application. Also called Total Exchange Capacity, Base Exchange Capacity, and Cation Adsorption Capacity.

Cation Exchange Resin. Highly polymerized synthetic organic compound which consists of a large, non-diffusible anion and a simple, diffusible cation, which later can get exchanged for a cation in the medium in which the resin is kept.

Cattle. Bovine animals. There are 250 major breeds and nearly 1000 breeds worldwide. They fall into two groups, those mainly European breeds developed from *Bos taurus,* and those developed from *Box indicus* (Indian cattle or Zebus). Cattle breeds are classified as beef, dairy or dual-purpose.

Cattle Filarid. A widely distributed nematode, found in the abdominal cavity of cattle.

Cattle Grid. A type of grate comprising parallel bars in a frame covering a pit in the road. Usually used to replace gates. The spaces between the bars are wide enough to discourage stock from crossing but are not a hindrance to humans or vehicular traffic.

Cattle Grubs. Maggots which are found on livestock. They can be controlled by hand removal or by tracting with rotenone containing materials such as benzol, iodoform ointment etc.

Cattle Lice. Two types of cattle lice have been recognised. The suctorial lice is able to pierce the skin of the animal and suck its blood. The biting species have got chewing mouth parts which make them to feed directly upon skin tissues. Both species attach their eggs firmly to the hairs of cattle. Cattle lice could be conveniently controlled by dipping, e.g., nicotine dip, arsenic dip. A sprayer or duster may also be used.

Cattle Lifter. A stealer of cattle.

Cattle Plague. A viral disease of cattle characterised by inflammation and ulcerations of mucous membranes. Eradicated from Britain in 1877. Still seriously affects cattle in Asia and Africa. Also called Rinderpest or Bovine Typhus.

Cattle Scab. Contagious skin disease which is affecting cattle of all classes and ages and is caused by mites which are parasites obtaining their food from the tissues of the host animal. It can be eradicated by dipping or spraying lime-sulphur dip, petroleum dip etc.

Cattle Tick Fever. An infectious disease of the blood of cattle which is caused by the development and actively of minute protozon parasites which are conveyed to the animals by the cattle-fever tick. The disease has been characterzied by high fever, destruction of red corpuscles, enlarged spleen, enlarged liver, thick and flaky bile, bloody urine. The eradication of the ticks is done by killing them on the pasture and on the cattle. Animals could be immunized by introducing the casual parasite into the system by blood inoculation.

Caul. A membranous sac, part of the amnion, enclosing the head of an animal at birth.

Caustic Paste. Prepared from one part arsenic trioxide mixed with five parts flour and wet with water to form a paste. This paste finds use in the treatment of summer sores of horses.

Cave Biology. Study of cave-dwelling organisms (troglobionts).

Cavie. A hen-coop or cage.

Cavings. The small pieces of straw and leaves broken up by a Threshing Machine and delivered from it after shaking out the grain and chaff.

Cave Biology. Study of cave-dwelling organisms (troglobionts).

Cavy (Plural, Cavies). Refers to one of the several species of tailless or short-tailed rodents of the family Caviidae, like the guinea pig, capybara and mara.

Cayenne Tick. Parasite which is found on horse. Clipping helps to keep them off horses.

Cecidiology. Division of biology which deals with the study of galls on plants caused by insects, mites and fungi.

Cell Wall. The limiting layer of a plant cell enclosing the protoplasm, comprising the plasma membrane and cellulose laid down as a crystal lattice.

Cellulose. A fibrous polysaccharide, the fundamental constituent of the Cell Walls in plants. (Carbohydrates)

Cemented Soil. Soil in which the grains or aggregates adhere firmly and are bound together by some material which serves as a cementing agent such as colloidal clay iron silica, or alumina hydrates, lime etc.

Central Leader. In this system a plant or a tree has been trained to form a trunk which extends from the surface of the soil to the top of the tree.

Central Seed Laboratory. Under Sub-section (i) of Section 3 and of the Indian Seeds Act, 1966; the Seed Testing Laboratory at IARI New Delhi has been recognised as the Cental Seed Testing Laboratory.

Centrifugal Divider. An instrument which is used for dividing a sample in the Seed Testing Laboratory. It uses centrifugal force to mix and scatter seeds over the dividing surface.

Cereal Forage. Cereal crop which has been harvested when immature for either hay, silage, green feed or as pasturage, *e.g.*, Bajri fodder.

Cereal(s). Cultivated members of the grass family whose seeds or grain are used to provide flour for breadmaking or as animal feed. The grain is rich in Starch but also contain valuable Proteins and Vitamins. The main cereals grown in India are Barley, Maize, Oats, Rye and Wheat.

Cereal Substitutes. Substances used by animal feed compounders to provide the carbohydrate content of the feed instead of traditional cereals. The common substitutes include manioc, maize gluten, rice bran, milling offals, grain screenings, citrus and beet pulp, and brewing and distillery waste.

Certification Agency. An agency which has been established under Section 8 or recognised under Section 18 of the Indian Seeds Act, 1966. Its function has been to certify seeds of any notified variety.

Certification Sample (Seed). Refers to a smaple of seeds which has been drawn by a certification agency or by a duty authorised representative of the agency. The sample would be for the purpose of determining if the lot of seed from which the sample was collected is having statisfactory germination and purity to get tagged and sold as certified seed.

Certification (Seed). Refers to the system of maintaining the genetic purity and quality of seeds. The crops offered for certificatin are grown as per requirement of seed certification standards which have been established by a seed certification agency. The crop has been inspected several times to ensure purity and quality of seeds.

Certification Tag. Lobel of specified design which indicates the certificate issued by the seed certification agency.

Certified Seed. Seed which is grown from certified seed stocks and that which is meeting the standards set up by a certifying agency for germination, presence of weed or other seed and impurities, freedom from disease-carrying germs.

Certified Seed Producer. Person growing or distributing certified seed as per procedures and standards of the certification agency.

Certified Seed Sample. Samples that have to be submitted by a seed certification officer or inspector for escertaining the quality to be tagged and sold as certified.

Cesspit, Cesspool. A pit, well or pool into which liquid wastes and manure are directed. The solids are left behind as the liquid drains away.

Chafers. Various beetles, the grubs of which are pests in old grasslands, usually those which are in slightly dry situations.

Chaff. In general terms any worthless matter. The husks of corn separated from the grain during. Threshing or Winnowing. Also short lengths of cut hay or straw. Also known as dowse or chop.

Chaff Cutter. Machine used for chopping fodder.

Chaffing. The chopping up of straw and hay for feeding to stock. Chaff is commonly mixed with concentrate rations for horses. It discourages them from bolting the ration and compels them to chew it.

Chain. A measure of length equivalent to 22 yds (20.1 m) or 100 links.

Chain Harness. A type of harness used to draw ploughs or other implements having no shafts. Also known as sling gear.

Chain Harrow. A type of Harrow, lacking a rigid frame, comprising a series of flexible chain links bearing spikes or knifetines, generally double-ended and reversible. This arrangement permits the harrow to follow uneven ground very closely.

Chalaza. 1. The spiral albuminous band anchoring the yolk at each end to the shell in a bird;s egg. 2. The base of the ovule in plants where the funicle or ovule stalk is primarily attached.

Chaldron. A measure equal to 4 quarters or 32 bushels.

Chalk. A soft white or greyish, very porous type of limestone, very rich in calcium carbonate. It is spread on the land to correct soil acidity and calçium deficiency, either as rough chalk or after crushing (ground chalk) when it is more effective. It is sometimes klin dried, finely ground, and sold as ground carbonate of lime.

Challenge Feeding. The feeding of diary cows in early lactation with concentrates at a rate in excess of standard recommendations, to provide extra energy which would otherwise be derived by the breakdown of body tissue. Also called lead feeding.

Channel Terrace. Refers to the type of contour terrace for soil conservation, wide and relatively shallow channel excavated at suitable intervals on a falling contour with a suitable longitudinal gradients excavated soil deposited as a wide, low ridge along the lower ridge of the channel. It is recommended for relatively impervious soil and in heavy rainfall tracts.

Chapped Teats. Caused by any irritation like sudden chilling after sucking by the calf, wet milking, wet bedding or freezing in water. Treatment includes washing the udder with warm soapy water, rising and then drying with a towel and painting, the chapped surface with glycerine iodine solution.

Character. 1. The nature and way of behaviour of an animal. 2. A specific recognişable feature transmitted from generation to generation, *e.g.*, height, colour, etc. In plant and animal breeding attempts are made to breed-In or breed out characters considered desirable or otherwise.

Chase. The hunting of wild animals by pursuit. Also an open area or preserve for game. Beasts of chase are wild animals generally hunted, e.g., buck, doe, fox, marten and roe.

Chats. The smallest potatoes in a crop, either too small or of inadequate quality to be classed as ware or seed potatoes.

Check Dam. Small low fixed dam which has been constructed of brush, logs, timber, loose rock masonry or concrete, in an eroded channel for reducing the slope of the water flowing therein during high stages, and also the resulting velocity, thereby preventing excessive scour and erosion and inducing deposition.

Checking. The tilting of a plough to the land. (Boarding).

Check Irrigation. Method of irrigation in which the water has been applied to a field or orchard which is divided into a series of checks by earth ridges, the water flowing from one check to another along the slope.

Check Method of Irrigation. Method of irrigation by which water has been run into relatively levelled plots which are surrounded by small ridges. It is the most common method in India and is suited to irrigation of grain and fodder crops in heavy soils where water is absorbed very slowly and so must stand for relatively long time to assure adequate penetration.

Check Row Planting. Refers to the method of planting in which row to row and plant distance has been uniform. In this method, seeds have been planted precisely along straight parallel furrows. The rows are always in two perpendicular directions. A machine used for check row planting is known as *check row planter*.

Cheek. One of two corresponding sides of an implement. Often used for the side of a plough facing the unploughed part of a field.

Cheese. A nutritious foodstuff prepared from the curd of milk, coagulated by rennet or acid, separated from the whey, and pressed into a solid mass. The nature of the Butterfat content of the milk is important. Small fat globules (characteristic of Ayrshire and Friesian milk) which rise extremely slowly, are required to produce an even texture of the cheese.

Chemical Analysis. The determination of the composition of a substance using chemicals to break it down into simpler substances. The analysis may be qualitative, showing only the constitutents present, or quantitative, showing the actual amounts present. Used, for example, to determine the Fertilizer requirement of soils, or the compositional quality of milk.

Chemical Score. Refers to a method which is used for determining the protein quality of feed in which the content of each essential amino acid has been determined and expressed as the percentage of standard, the lowest percentage is considered to be the score.

Chemical Weed Control. The application of herbicides to crops on bare land to kill weeds.

Chemotroph. An organism (bacterial) that uses chemical energy for the synthesis of food.

Chenopodium Oil. Volatile oil which is obtained from the above-ground parts of the American wormseed plant. It is a pale yellow liquid and is used as an anthelmintic. It may be used for the treatment of ascariasis in calves but is not given to calves suffering from diarrhoea or constipation.

Chernozem. Soil characteristic of some grassland vegetation in warm areas having moderate rainfall; dark in colour because of a high content of organic matter.

Chersophyte. (GK *chersa,* dry places; *phyton, plant*) A plant which grows on waste land or on shallow soil.

Chestnut. 1. A horny knob on the inside of a horse's forelegs. 2. A reddish-brown colour, used descriptively of horses. 3. A hardwood tree *(Castaned sp.).* The Sweet Chestnut *(C. Sativa)* is grown as a coppice crop, particularly in South-East England, much used for hop poles and fencing.

Chestnut Soil. Zonal group of soil which are having a dark brown surface horizon which grades below into lighter coloured soil and finally into a horizon of lime accumulation developed under mixed tall and short grasses in a temperate to cool and subhumid to semi-arid climate. They are lound on the arid side of the chernozem soils into which they grade.

Chewing Cattle-Louse. The only biting species of cattle lice which concentrates on the withers and around the root of the tail.

Chewing the Cud. An activity of Ruminants, involving the mastication for the second time of food which has previously been swallowed and passed in to the first stomach or Rumen.

Chick. A young bird about to be hatched or newly hatched.

Chicken. The young of the domestic fowl. Also a term for the flesh of a domestic fowl, not always very young.

Chifox. A breed of long-haired rabbit, variously coloured.

Chigger. Mite which is able to feed for one or two days on the blood serum from the host, bringing about the itching and redness of the skin before they drop off to move on the ground. They affect both man and animals.

Chilling Requirement. Cold period which is required by certain plants and plant parts so as to break physiological dormancy or rest. The chilling requirement is generally expressed in terms of the required number of hours at 7°C or less.

Chimeras. Plant, or part of a plant which is composed of two or more genetically distinct tissues and is growing adjacent to each other as parts of the composite plant. Plants having variegated foliage, as in coleus, dahlia, citrus, vitis, chrysanthemum are examples of chimeras.

Chip. A fruit basket made from interwoven thin strips of wood.

Chip Budding. Method of budding which can be carried out at times when the bark is not slipping. It involves removing a chip of bark from a smooth place between nodes near the base of the stock and replacing it with another chip of the same size and shape from the budstick having a bud of desired variety. The chips in both stock and the budstick should be cut in the same manner. The chip bud is then wrapped to seal the cut edges and to hold the bud piece tightly into the stock.

Chirping. Production of sound by an insect, generally by rubbing one part against another.

Chisel Plough. A type of very heavy Cultivator, with large section tines, the points of which incline forward, and are drawn through the soil at a depth greater than in normal ploughing. The underlying layers are burst without subsoil being brought to the surface.

Chit. A shoot or sprout.

Chitin. A nitrogenous polysaccharide (Carbohydrates) found in the exoskeleton of insects and in the cell walls of many fungi. It provides mechanical strength and resistance to chemicals.

Chitinase. An enzyme which hydrolyses chitin.

Chitterlings. The smaller intestines of a pig or other animals sometimes used for food.

Chitting House. A building in which trays of potatoes are stored in stacks during the winter and stimulated to sprout before planting, by the provision of controlled heating to prevent frost damage and regulate the rate of growth, and movable fluorescent lighting.

Chlordane. A persistent, highly toxic, organochlorine insecticide, used to control earthworms in turf.

Chlorinated Hydrocarbons. A group of insecticides which mainly act on the central nervous system. The tend to be persistent in the environment and can build up of toxic levels in the body fats of other animals higher up food chains. They include D.D.T., Aldrin, Dieldrin and Endrin amongst other. (Organo Phosphorus Compounds).

Chlorine. Plant food element which has been absorbed by plants as chloride ($C1^-$) ion, exact role in plant nutrition not known. Chlorine compounds used as disinfectants and deodourants.

Chlorine Solution. Non-irritating disinfectant. It is used for washing udders of cows as well as teat cups of miling machines by dipping for one minute in a solution having 200-300 ppm.

Chlorine Test (Live Stock). This test finds use for determining mastitis. The test is based on the fact that normal milk is having a chlorine content varying from 0.08% to 0.14% and that the chlorides of the blood escape into the milk and the chlorine content of the milk has been raised with the inflammation of udder. It is a rapid test which should be run by a dairyman, but it should be used with a second test for confirmation of the findings.

Chlorobutanol. It forms colourless to white crystals having camphoraceous odour which is commonly used as antiseptic, hyponotic and sedative. Heavy feeding of chlorobutanol may bring about paralysis and sometime even death of the animal.

Chlorocruorin. Green respiratory pigment which occurs in blood plasma of certain worms.

Chloromycetin. An antibiotically active chemical compound ($C_{11}H_{12}NO_5Cl_2$) which is produced by *Streptomycer venezuelae* during the fermentation of a suitable culture medium.

Chloropyll. The green pigment found in plants essential to the process of Photosynthesis.

Chloropicrin. (Tear Gas). Chemical which is used for soil fumigation. It is a liquid which has been applied with an injector. Chloropicrin is effective against nematodes, some weed seeds and insects.

Chloroplast. A chlorophyll containing organelle in a plant cell, the site of photosynthesis.

Chlorosis. A condition of plants characterised by normally green parts becoming pale green or yellow (chlorotic). It is due to the prevention of Chlorophyll formation, caused by various factors including lack of light, magnesium or iron deficiency, and excess calcium (when called Lime chlorosis). Also called green sickness.

Chocolate Spot. A fungi disease (*Botrytis cinerea* and *B. fabae*) of beans, particularly when winter sown. It causes the development of chocolate coloured spots on the leaves, often quite rapidly, in warm wet conditions following late spring frosts. The shoots become black and die.

Choke. A fungal disease (*Epichloe typhina*) of cocksfoot, timothy and other grasses, characterised by fungal cylinders around the stems and leaves, affecting seed production.

Chopper Spreader. A mechanism sometimes attaches to the straw outlet of a Combine Harvester which chops the straw into short lengths and spreads it over the field.

Chopping out. The removal of unwanted plants in a crop.

Chorion. The outermost membrane around a mammalian foetus (contiguous with the uterus surface), and around a developing chick in an egg.

Chromosome. A thread-shaped body, comprising mainly D.N.A. and protein, carrying the Genes, and present in the nucleus of a cell. (Diploid, Haploid, Meiosis, Mitosis)

Chrysalis. A Pupa.

Chunk Honey. Honey containing at least one piece of Comb Honey.

Churn. 1. A machine which agitates cream or whole milk turning it into butter. 2. A large metal milk container or can, commonly of

10 gallons capacity (occasionally 12 gallons), with a close-fitting, muschroom type lid.

Circling Disease. Specific bacterial disease which effects mammals including man and birds. It is characterized by meaningo-encephalities and sometimes abortion in domestic ruminants and human beings. It is caused by *Listeria monocytogenes.*

Citrate Soluble. Refers to that part of the total phosphoric acid in a fertilizer which is insoluble in water but soluble in a neutral solution of citrate of ammonia. It is mostly in the form of dicalcium phosphate, $CaHPO_4$.

Citrus Molasses. Product which is obtained from the juice of citurs wastes.

Clasper. One of a pair of appendages at the tail end of a caterpillar by which it clings to leaves, twigs and other surfaces.

Clat. To remove wool from around the udder and the inside of the thighs of a ewe prior to lambing.

Clay. A constituent of Soil comprising extremely finegrained particles, less than 0.002 mm in diameter, characterized by high moisture retention. Sticky and plastic when wet, hardening and shrinking on drying. (Clay Minerals, Clay Soils, Loam, Sand, Silt).

Clay Minerals. The particles comprising clay consist of layers of hydrated aluminium and magnesium silicates forming a crystal lattice. Various types of clay mineral exist, *e.g,* kaolinite and montmorillonite, having different layer structures, and differing abilities to absorb and retain water, and to absorb and exchange cations (Cation Exchange). The agricultural productivity of a soil is related to the nature and content of the clay minerals in it and to its ability to retain (and subsequently release for crop use) cations, particularly those applied in the form of fertilizers.

Clay Soils. Soils containing large amounts of clay, often called heavy soils because they require much more effort (*i.e.*, greater tractor power) to plough and cultivate than sandy or light soils. They are sticky and swell when wet and are liable to poaching, particularly in winter. They retain moisture well but become hard, shrink and crack in dry weather. Artificial drainage of clay soil is often carried out, particularly for arable cropping. (Land Drainage, Loam).

Clay Loam. Heavy soil intermediate in texture between clay and loam.

Clay Pan. (1) In common usage, it refers to a dense, heavy subsoil, more or less impermeable to roots (2) A dense and heavy soil horizon which underlies the upper part of the soil profile, hard when dry and stiff or plastic when wet.

Clay Pots. Familiar red clay flower pots, which are long used fror growing young plants.

Clean Cattle. Cattle which have not been used for breeding, *i.e.*, maiden Heifers, steers and young Bulls.

Clean Corn. A sample of wheat, barley, etc., not mixed with the seeds of other plants.

Clean Cultivation. Periodic soil tillage which carried out to eliminate all vegetation other than the crop being grown.

Clean Cultivation. Cultivation which is carried out to prevent the growth of all vegetation except the particular crop desired.

Clean Land. Land free from weeds.

Clean Legged. A descriptive term for a horse with few or no long hairs (known as feather) on its legs.

'Clean' Pasture. The sward after an arable crop, following reseeding, ungrazed by livestock and therefore presumed to be uncontaminated by parasitic worm larvae.

Clean Tillage. Periodic soil tillage which is carried out to eliminate all vegetation other than the crop being grown.

Cleft Graft. A graft involving the inseration of the scion into a wedge-shaped cut in the centre of the stock.

Cleg. One of various species of blood-sucking, two-winged, tabanid flies (*Haematopota sp.*). They are ashy-grey with finely mottled wings and large bright eyes and are troublesome to horses. Also called horse-fly.

Cleft Palate. Hereditary defect in pigs. It is characterized by their inability to nurse, they die soon after birth.

Cleistogamy. Close fertilization within the unopened flower due to self-pollination.

Cling. Diarrhoea in animals, particularly sheep.

Clip. 1. To remove or Shear the wool from a sheep. 2. The wool removed from either a single sheep or a whole flock. 3. One of several projections which prevent a horseshoe from shifting on a horse's foot.

Clod. Mass of soil which is produced by ploughing or digging and usually slaken easily with repeated wetting and dying in contrast to a ped which is a natural soil aggregate.

Clonal Selection. Refers a method of selection of desirable clones from the mixed population of a vegetatively propagated crop.

Clone. A group of plants or animals all having an identical genetical make-up, being the descendants of a single parent. In plant breeding a clonal stock of plants can be vegetatively raised by taking cuttings from a single parent plant.

Closed Formula Mixed Fertilisers. The fertiliser grade has been disclosed on each bag of such fertiliser mixture but the ingredients or straight fertilisers used in formulating the mixture have not been disclosed. In India, fertiliser mixtures sold to the cultivators are usually of the closed formula type.

Cloven Hoofed. A descriptive term of animals with a divided hoof such as cattle and sheep.

Clover. A large genus (*Trifolium*) of leguminous plants- characterized by trefoil leaves, having small flowers with tight roundish heads, and producing one-to four-seeded pods. Root nodules contain bacteria capable of nitrogen fixation. A number of species are grown for fodder or are sown with a mixture of grass seeds to produce Leys, *e.g.,* Aslike Clover (*T. hybridum,*) Wild White Clover or Kentish Clover (*T. repens*), Purple or Red Clover (*T. pratense*). Clovers usefully knit the sward together and help keep out weeds, especially in longer leys.

Clover Hay. Hay derived from a clover crop. Also hay from a temporary Ley as opposed to that derived from permanent grassland (Meadow Hay). Also knwon as seeds hay.

Clover Rot. A fungal disease (*Sclerotinia trifoliorum*) of clovers and related plants, characterized in its early stages by leaves being covered with small brown spots. Plants later wither and die.

Clover Sickness. A condition of a field on which clover has been grown for too long and which is heavily infected with Eleworms and fungus diseases.

Club Root. A fungal disease (*Plasmodiophora brassicae*) of cruciferous plants, particularly swedes, turnip, rape, cabbage and other Brassicas, which causes swelling and distortion of the roots. Growth gets stunted and leaves become pale green. The spores can remain alive in the soil for several years and are particularly active in acid and wet soil conditions. Also known as Finger-and-Toe.

Cluster. 1. The four teat cups of a Milking Machine applied to a cow's udder. 2. A swarm of bees after settling. 3. The leaves, flowers or fruits growing from a common point on the stem or shoot of a plant.

Clutch. A batch of eggs laid by a bird. Also the brood of chicks hatched from such eggs.

Coated Seed. When seeds are first treated with strong water stable adhesives, they bind to the seed coat and adhere particulate matter of various kinds tightly to the seed to better withstand the hazards of sowing.

Cobalt (Co). A metallic Trace Element present in small amounts in soil and herbage, required for the assimilation of iron into haemoglobin in red blood corpuscles. When deficient it causes pining' in sheep.

Cobweb Theorem. An economic theorem that producers of a commodity expect the price in the next period of production to be the same as current prices and thus plan production accordingly. It is most appropriate in the case of non-storable commodities.

Cocaine. An alkaloid obtained from the leaves of *Erythroxylum coca.* It finds use as a local anesthetic and mydriatic. Its solution in water is sometimes applied to soothe the pain in eyes.

Coccidioidal Granuloma. Refers to the non-contageous fungus disease of animals, which is characterized primarily by multiple pulmonary granulomas and dissemination to other organs. *Coccidioides immitis* is the casual organism.

Coccidiosis. An intestinal disease of livestock and poultry caused by various microscopic protozoan parasites of two genera *Eimeria sp.* and *Isospora sp.,* belonging to the Order Coccidia. Characterized by diarrhoea and emaciation.

Coccus. 1. A spherical bacterium, as distinct from rod-shaped or vibrio (spiral-shaped) bacteria. 2. A portion of dry fruit contain ing one seed when it breaks up.

Cock 1. A general term for a male bird. For poultry, usually applied to males over 18 months old. 2. Slang for woodcock. 3. A small pile of dung or of a cut crop drying in the field.

Cock Hop. A freak hop Aone bearing a single or several small leaves between the bracts.

Cockerel. A young male chicken, usually applied to one less than 18 months old. Also a male turkey less than 12 months old.

Cocking. The forming of heaps or cocks of a loose cut crop for drying in the field.

Cockle. A shrunken seed, badly formed ; crinkled or puckered.

Cocksfoot. A perennial grass (*Dactylis glomerata*), with characteristic one-sided flower-heads of rounded or oval individual flower spikelets on wiry branches, and dull green to deep blue leaves. Deep rooting, very high yielding used extensively for pasture.

Coconut Cake. A concentrate feeding stuff reasonably rich in Protein, used frequently in rations for dairy cows, fattening cattle and pigs. It produces a harder fat than most other oil seed products. Not as readily available as it used to be.

Cocoon. A protective silky sheath spun by many insect larvae (*e.g.*, the Caterpillars of butterfiles and moths) in which they pupate (Puppa). Also a similar sheath or capsule in which spiders and earthworms lay their eggs.

Codlin(g). A type of elongated apple.

Codlin(g) Moth. A small moth the caterpillars of which eat the inside of growing apples and cause them to fall prematurely.

Cod-liver Oil. The oil obtained from the livers of codfish, rich in vitamins A and D, valued for feeding to young animals to improve condition. It is particularly helpful in preventing rickets developing.

Coggin test. A test used to detect anaemia in horses.

Coir Yarn. String or twine made of coconut fibre on which bines are trained in hop gardens.

Coffin-Joint. Joint located between the second and third phalangel bones of the hoof.

Coital Vesicular Exanthema. An acute contageous disease of cattle characterized by fever, rhinitis, tracheits and considerable loss of condition and milk yield.

Cold Blooded. Poikilothermic ; lacking ability to regulate the body temperature by physiologic means; commonly said of fish, reptiles, and amphibians whose temperatures approximate that to the environment.

Cold Branding. Refers to the branding of farm animals by using caustic depilatories. This produces a permanent skin brand without discomfort to the animals.

Cold Brooding. Also known as hay-box brooding. It is a simple method of rearing of chicks when it is not too cold, by employing an ordinary basket or small box lined thickly with fine hay or old flannel strips of feathers and provided with small hole or outlet to allow passage to chicks.

Cold Storage. The preservation of perishable produce (*e.g.*, apples, meat, etc.), usually in bulk, in refrigerated chambers. The low temperature conditions inhibit the multiplication of bacteria and fungal sproes, etc.

Cold Text. A germination test in which the seeds are planted for a period in cool, moist unsterilized soil before they get transferred to a higher temperature. This is done for studying the effect of possible unfavourable planting conditions while determining seedling vigour.

Colibacillosis. Disease which is caused by *Excherichia coil*. It is found in the intestine, where it lives in symbiosis.

Colic. Abdominal pain and associated symptoms which results from any one numerous disordres of the abdominal cavity.

Coli Septicaemia. A bacterial infection (*Escherichia coil*) of poultry. It is usually secondary to other respiratory diseases and is characterized by inflammation of the air sacs which often contain purulent material. Young birds between 6 and 10 weeks of age are particularly susceptible, mainly in winter in conditions of poor ventilation or overcrowding.

Coliform Bacteria. Those bacteria commonly present in animal intestines and passed out in faeces (*e.g. Escherichia coli*). The presene of such bacteria in milk indicates faecal contamination.

Coliform Test. Test which is used for estimating coliform bacteria in a sample of milk. The presence of such bacteria is an index of unhygienic method of milk production.

Colitis. Inflammation of the large intestine which is often caused due to diarrhoea, constipation, etc.

Collagen. A proteinous substance, the principal fibrous constituent of skin, tendon, ligament and bone. When boiled it produces gelatin.

Collective Fruit. Single fruit which is formed from several flowers.

Colony Brooders. Brooders which are used for rearing chicks in lots from about 60 to a few hundred, with kerosene oil, coal, wood, oil, gas or electricity supplying the fuel.

Colostral Milk. First milk which is secreted soon after parturition. Colostral milk helps in cleaning out the digestive tract and provides the newborn animal, resistance against harmful bacteria.

Colostrum. The milk given by cows and other mammals during the first few days after giving birth. It is particularly rich in proteins including Antibodies, and has a higher fat content than normal milk. Also called beestings or beistings.

Coloury. A descriptive term applied to dried hops having a distinctive green tint.

Colt. A young male horse less than four years old (five years if thoroughbred).

Colza. The oil yielded by Rape seeds (also called cole seeds).

Comb. 1. A fleshy crest on the head of some birds. 2. The cellular structure constructed of beeswax in hives in which bees store surplus honey. In commercial beekeeping hives contain frames holding a thin sheet of wax on which bees construct the six-sided cells which form the comb. Also called honeycomb. 3. The lower blade of the cutting section of a powered sheep- shearing handpiece. 4. The edge at the top of a Furrow slice. Also called arris.

Comb Honey. Honey stored by bees in the cells of freshly built broodless combs.

Combine Drill. A Drill which sows seeds and fertilizer simultaneously, usually down the same seed-tube, although there are separate hoppers for seed and fertilizer. Combine drills are difficult to clean out effectively.

Combiner. Machine which undertakes the functions of a reaper, thresher and winnower. Its main functions are as follows : (*i*) Cutting the standing crop (*ii*) Feeding the cut crops to threshing unit (*iii*) Threshing the crop (*iv*) Cleaning the grains from straw and (*v*) Collecting the grains in a combiner.

Commercial Feed. Materials distributed for use as feed or for mixing in feed for animals other than man.

Common Stomach Worm. Nematode found in the fourth stomach or cattle, goat and sheep. It is 1 to 3 cm long and as thick as an ordinary pin. This blood sucking parasite is responsible for the loss of flesh, general weakness, anaemia, and diarrhoea. Infested animals may be treated with copper sulphate or phenothiazine.

Community. 1. A term applied to any assemblage of plants making up a distinct vegetation type, e.g., deciduous woodland. 2. A short-term for the European Economic Community.

Compacted Surface. Compacted surface or crust on soil which is somewhat loose under-neath but does not break regularly during ploughing.

Companion Crops. Two crops grown together in the same field, one of the crops, and often both, benefiting from the presence of the other. The term is particularly applied to crops grown together for seed production which have similar cultivation and management needs and whose seed can be separated afrter harvesting (e.g., clover and ryegrass, the clover facilitating soil Nitrogen Fixation, beneficial to ryegrass growth, whilst the ryegrass assists in the harvesting of the clover).

Compensatory Amounts. Monetary Compensatory Amount.

Compensatory Growth. The phenomenon of animals turned out to pasture after a period of energy restriction exhibiting greater growth rate than animals which have not been so restricted. It is due to the greater feed intake following the period of energy restriction.

Competition. Demand of organisms within a community for the same substance. Plants may compete for light and nutrients; animals for food and shelter. Competition usually causes the displacement of one organism by another.

Complement. A constituent of blood serum, the presence of which is required for an Antibody to kill an Antigen.

Complement Fixation Test. A blood test used to diagnose certain bacterial diseases, e.g, Brucellosis.

Complete Diet. Winter feed for livestock comprising a mixture of Concentrates and bulk forage fed as a complete mixture, the animals usually receiving no other feed.

Complex Fertilisers. Commercial fertilizers having at least two or more of the primary essential nutrients. When such fertilisers have only two of the primary nutrients they are known as **Incomplete Complex Fertilizers.**

Composite Seed Sample. Seed sample composed of mixture of different subsamples taken from various parts of a seed lot

Compost. A manure derived from decomposed plant remains, usually made by fermenting waste plant material (e.g., straw, mowings, etc.) in heaps, usually in alternate layers with added lime, nitrogen and water. Compost is usefully used in greenhouses, to enrich soil, either dug in or as a surface Mulch. Straw compost is valuable in building up poor soils for intensive vegetable cropping and assists in maintaining soil structure on heavier land.

Compost Activator, Compost Accelerator. A chemical substance which promotes decomposition and fermentation of decaying plant remains in a compost heap.

Compound Feed. Animal feed composed of several different feedingstuffs (and including major minerals, Trace Elements, Vitamins and other additives) in proportions providing a balanced diet. The compound manufacturers provide a range of feeds suitably balanced for all types of stock at all growth and development stages.

Compound Fertilizer. A fertilizer containing a mixture of two or three of the major plant nutrients (i.e. Nitrogen, Phosphorus and Potassium). Many proprietary compound fertilizers are produced with differing nutrient ratios to suit specific crops and soils. Mostly produced in easily stored granulated form, but powders and less concentrated liquids are available. Also called mixed fertilizer.

Compound Leaf. One divided into separate leaflets, e.g., Clover. (Simple Leaf)

Compressed Pockets. Hop Pockets compressed to a high density for storage or export.

Compressibility—of Soil. The susceptibility of a soil to compact under a heavy load (i.e. a tractor).

Concave. A semi-circular grating in a Threshing Machine, encirclind the drum, against which the grain is dislodged by the rotating Beaters on the drum.

Concentrates. A term for a variety of animal feedingstuffs with a high food vfalue relative to volume, with low fibre content, some rich in Protein, others rich in Carbohydrates or Fat. Concentrates include cereal grains and their by-products, leguminous seeds, oil seeds, oil Cakes and Meals, and various animals by-products.

Mainly supplied by compound feedingstuff manufactuers as expertly balanced formulations, often computer assisted, for all types of farm stock. Ingredients may vary in relation to current prices and availability and Trace Elements, Vitamins and Minerals are usually included to produce complete foods. Some farms mix their own concentrates using home-grown cereals.

Condensed Milk. Enriched milk produced by evaporating much of its water content, and by the addition of sugar.

Conditioner (of Fertilisers). Material which is added to a fertiliser to prevent caking and to keep it free-flowing.

Cone. 1. The fruit of the female hop-plant, growing as a cluster of hops on a Bine. 2. The fruit of a coniferous tree, consiting of compact over- lapping scales, arranged spirally around a central axis.

Confiers. A group of trees and shrubs with needle-like leaves and typically bearing cones, e.g., pine, fir, spruce. Coniferous trees or softwoods grow quickly and provide early profit where grown commercialiy. They are especially suited to exposed hili and acid conditions (Broadleaf)

Conservation. 1. The optimum rational use of natural resources and the environment, having regard to the various demands made upon them and the need to safeguard and maintain them for the future. 2. The protection of the soil against erosion or loss of fertility. 3. The preservation of grass as Hay and Silage and of fodder crops for winter feed. Also applied to the preservation of certain growing crops *in situ* for later use, e.g., frost- hardy Thousand Headed Kale kept through the winter for folding in spring.

Consociation. Climax plant community which is characterized by a single species.

Constipation. The term used for the retention of faeces in the intestines because of difficult evacuation. It may develop due to feeding of coarse, dry, or indigestible feed or due to over-feeding or from overeating.

Consumptive Water Requirement. The amount of water potentially which is needed to meet evapo-transpiration needs of the plant so that the plant does not suffer in its growth through short supply of water.

Contact Animal. An animal which has been in contact with a diseased animal and may need to be isolated if the disease is infectious.

Contact Herbicides. Herbicides which kill only those parts of a plant with which they come into contact and used mostly to control annual weeds when seedlings. They have very little residual effect and are normally applied just before crop sowing.

Contact Placement. Drilling of seed and fertilizer together while sowing.

Contagious. Contagious disease which gets conveyed from a sick animal to another by actual contact.

Contaminants. Those factors affecting genetic and physical quality of the seed. For example of types, foreign pollen, other crop plants, weed plants, plants affected by designated diseases, weed seeds etc., are regarded as contaminants.

Continuous Brooders. Brooders which are used for rearing chicks in large number with coal, gas or oil suppling fuel, used in big commercial enterprises.

Continuous Drying. Refers to the process of drying with a continuous flow of grain and air, in contrast to back operation.

Continuous Layerage. Refers to an asexual reproductive process of plant propagation in which the whole branch except the tip has been buried 8-10 cms of earth and allowed to develop roots while still attached to parent.

Continous Ploughing. In this method, first the headland gets marked and the first ridge is set up at three quarter of land width from the side. The outer ridges are set at full width over the field. The operator starts ploughing between the first ridge and the sideland, continue to turn left and cast in the three quarter land until a quarter land width of ploughing gets completed on each side.

Contour Border Irrigation. Refers to the method of irrigating gently sloping fields, the whole area being divided into strips by ridges along the contours and cross ridges. The ridges confine the water to a particular strip till it is completely full, before, it is allowed to flow to the next lower strip. Also called contour check irrigation.

Contour Cropping. Cultivation of crops along the contours of a slope.

Contour Farming. Refers to the method of cultivation wherein operations including sowing have been carried out along the contour. It reduces runoff, conserves more moisture for crop production, reduces soil losses and increases crop yield.

Contour Furrows. Furrows which are ploughed on the contour on pasture of range land to prevent soil loss and allow water to penetrate the soil. Sometimes used in planting trees or shrubs on the contour.

Contour Ploughing. Ploughing on the level following the contour, as distinct from up and down the slope. This practice is an effective method of soil erosion control.

Control, Biological. The employment of the enemies and diseases of a pest for the purpose of maintaining adequate control. The artificial application of biotic control.

Controlled Atmospheric Storage. Cold storage in which the concentrations of atmospheric gases have been adjusted to extend the storage life of fresh produce. Usually oxygen has been lowered and carbon dioxide raised.

Conveyors. Equipment which is used for moving the seeds in seed testing laboratory.

Cooperids. Small nematodes that are found in small intestine of cattle and occasionally of sheep and goat. They produce extensive intestinal inflammation in calves. Severely affected animals possess more or less resistant diarrhoea and become imaciated and anaemic as the disease progresses.

Copper. (Cu). A metallic chemical element, essential to animals in trace amounts, but toxic if consumed in quantity. Also required as a trace element by plants form various metabolic processes. Deficiency diseases may occur in plants and animals if it is not adequately available. (Swayback).

Copper Fungicides. Copper compounds which are used for killing fungi causing plant disease. They act by contact, killing only those spores or other parts of disease agents with which they are brought in to close contact, popular examples include copper fungicides like Bordeaux mixture; copper oxychloride, cuprous oxide, perenox perelan (a dusting powder based on copper).

Copper sulphate. ($CuSo_4$ $.5H_2O$). A blue crystalline salt used as a fungicide in bordeaux mixture. Also called bluestone or blue vitriol.

Coppice. A small wood of small trees and underwood cut periodically, essentially comprising sprouts from cut stumps.

Cordon. A fruit tree trained and closely pruned to grow as a single stem, supported by wire frames or walls.

Corm. A short, swollen underground stem, carrying buds, acting as an organ for storage and vegetative preproduction. Similar to a bulb but the food is stored in the swollen stem as distinct from the scales of a bulb. Examples, include crocus and gladiolus.

Corn. A general term for cereal crops (barley, maize, oats, rye and wheat) grown for their edible seeds or grain. The grain itself is also called corn. In America the term is restricted to maize. In Scotland and Ireland the term is generally restricted to oats, whilst in England it is mainly applied to wheat.

Corn Chandler. A retailer of grain.

Corn Cockle. A tall weed (*Agrostemma githago*) of cornfields. Hairy, long-stalked, with much divided leaves and single pale purple flower, with woolly sepals.

Carrecting Strip. An irregular strip or area of land which is laying between contour strips.

Corrosive Sublimate. Bichloride of mercury which is a violent poison. It has been highly effective as an antiseptic,, germicide, and disinfectant. As an antiseptic would wash and in dressings, it is used in 1:1000 to 1:5000 solutions; as disinfectant in 1:1000 to 1:5000 solutions. Corrosive sublimate solution has to be stored in wooden, glass or earthenware as it corrodes metals.

Cotton Boll. Fruit of ctoon plant is termed as boll.

Cottonseed Cake, Cottonseed Meal. An animal feedingstuff made from the seeds of the cotton plant after the extraction of oil, mainly used for adult ruminants. It has a high fibre content, is poor in calcium, with low protein quality. It is yellow-coloured due to the presence of gossypol which is toxic to young pigs and poultry, and is available either in decorticated or undecorticated forms.

Cotyledon. A seed leaf, forming part of the embryo of seeds. The first leaf to develop when a seed germinates, initially lacking chlorophyll. Monocotyledonous plants have one cotyledon in each seed whilst dicotyledonous plants have two types of plants are also distinguished by their leaves. The former possess long, narrow leaves with parallel veins (e.g., grasses) whilst the latter have broad leaves with branching xeins (e.g., broadleaf weeds such as dock, fat hen, etc.)

Coulter. The part of a plough which makes the vertical cut in the soil, usually mounted in front of the share, attached to the beam. There are three main types; knife, disc and skim coulters. Knife coulters were a feature of horse ploughs. Flat, circular, freely rotaing disc coulters are used on most tractor-drawn ploughs. They rotate as the plough moves, separating a slice of soil from unploughed land, which the share and mouldboard then turn to produce a clean furrow slice. Skim coulters are attached to the beam, either in front of knife coulters or behind disc coulters. Each acts like a small digger and turns a small furrow slice from the coulter of the main furrow into the furrow bottom, thus completely burying crop residues or manures also called colter.

Coulter Face. The side of a slice cut by a coulter ploughing, and then turned over by the mouldborad to become the front of the slice.

Counter Irritant. An agent causing a skin irritation which is intended to relieve another irritation of an organ below the skin. Mustard plaster is somtimes used in the treatment of pneumonia.

Counterfire. Intentionally burning a belt of trees towards an advancing forest fire so that both the fires may get extinguished when they meet.

Cover Crops. The crops grown primarily to cover the soil so as to reduce the lose of moisture due to leaching and erosion by wind and water. Many a times such crops are not harvested.

Cover, Ground. Any vegetation which produces a protecting mat on or just above the soil surface. In forestry, low-growing shrubs and herbaceous plants under the trees.

Cowpox. An acute virus disease of cattle, which is accompanied with a slight fever and a typical eruption which is usually confined to the teats and udder of the female or the scrotum of the male. The affected animals are promptly isolated and precautions taken to avoid carrying infection to other cows. Milking should be carried out gently and with as little discomfort to the animal as possible. Affected animal is to be batched with cresylic disinfectant like a 3% saponated cresol solution, and antiseptic ointment or oil applied.

Crabtree Effect. The continued improvement in the feed-use efficiency of an animal under feed restriction, so that the feed supply has to be repeatedly reduced further in order to maintain a flat growth rate.

Cracked Wool. A fleece which felts together and cracks open, especially in wet weather.

Cradle. A set of fingers or light rods in a light frame set into a scythe (a cradle-scythe), used to cut corn, laying it evenly with each stroke.

Craft. A class of work in agriculture such as milk production, machinery operation and maintenance, or plant nursery practice.

Crawler Tractor. Tractor which has endless chain or track in place of pneumatic wheels. This is also known as *track type tractor* or *chain type tractor*.

Cream. An oily substance containing fats which rise on milk left to stand. In manufacturing, milk is centrifuged to divide it into cream and skim milk. The cream is sold as liquid, clotted or sterilised, or churned into butter.

Cream Cheese. Soft cheese made from cream strained through muslin.

Cream Separator. A machine which uses centrifugal action to separate milk into lighter cream for buttermaking and Skim milk or separated milk containing the heavier particles. Either hand-operated or power-driven. Farmhouse buttermaking is now uneconomic except for home consumption, and cream separators are now rarely seen.

Creolin. Refers to an antiseptic, germicide, deodorant and disinfectant which is used for the control of certain external parasites of animals. It is a dark-brown liquid which consists of high boiling coal-tar, phenols and oils. It forms milky emulsion when diluted with water.

Crimp, 1. In general terms the waviness of wool. One of the waves in wool fibres. Measured as the number of crimps per unit length and an important quality in assessing manufacturing value. 2. To crush grass with a crimper.

Crimper. A machine with ribbed rollers between which grass is crushed during haymaking, allowing sap to be removed by the action of sun and air. Similar to a Roller Crusher.

Criss-Cross Breeding. A system of pig breeding in which the same two breeds are used as alternative sires on home-produced, cross-bred gilts. By comparison, rotational crossing involves the use of boars from three or four breeds used in strict rotation.

Crock. An old ewe or horse.

Critical State. Unique conditions of pressure, temperature and composition where in all properties of coexisting vapour and liquid become identical.

Critical temperature. Approximate temperature of the air when the heat lost by the animal body through physical means such as evaporation just balances the heat produced as a result of internal work and tissue oxidation, and for which further tissue oxidation is necessary to maintain the normal temperature of the body.

Critical Temperature for Water. Temperature at which the volume of water and the volume of saturated vapour formed from it are equal. This is about 365°C.

Critical Velocity. Of a liquid is that velocity of flow above which the flow ceases to streamline.

Crochets. Any small hook-like organs, particularly those on the prolegs of caterpillars.

Crooked Foot. A horse foot with one side wall higher than the other which is largely attributed to inherited faulty conformation with 'toeing out' or 'toeing in'. Sometimes it may also be caused by unequal pairing of the foot sole or by improper shoeing. Repeated trimming and use of special shoes are useful.

Crop. 1. Plants, carefully selected and developed over many years, sown on cultivated land to produce food for man and animals or raw materials, e.g., barley, potatoes. The term is also applied to plants which are not sown but come up naturally in cultivated land from wild seed, e.g., a crop of thistles. 2. To mow, cut, reap or gather a cultivated crop. 3. The total quantity produced, cut or harvested from cultivated land. 4. To bite off in eating, e.g., grazing sheep will crop a ley. 5. The expanded part of a bird's oesophagus or gullet, in which food is stored before being digested. Also called the craw. 6. A hunting whip with a loop as distinct from a lash. 7. An entire hide.

Crop Bound. A bird with its crop obstructed by material which cannot pass on to the stomach and gizzard, e.g. feather, straw, wool, etc.

Crop Cover. Subsidiary crop of low plants which has been introduced in a plantation to afford cover between or below the main crop; also any crop which is used to protect land from erosion. A cover plant is one suitable for use as a cover crop.

Crop Residue. That portion of a plant or crop which is left in the field after harvest, or that part of the crop which is not used domestically or sold commercially.

Crop Rotation. A definite succession of crops following one another in a specific order.

Crop Yield Index. Refers to a measure of comparison of the yields of all crops on a given farm with the average yield of these crops in the locality. The relationship has been expressed in percentage. It is a suitable method under Indian conditions where diversified type of arming has been followed.

Cropping Intensity. Refers to the number of crops which are raised during the year. It is measured as :

$$\text{Cropping intensity} = \frac{\text{Cropped area}}{\text{Sown area}} \times 1000$$

However, it does not take into consideration the length of growing period for various crops.

Cross-breeding. The mating of animals (or plants) of different breeds, but both of which are purebred, in order to combine the best characteristics of the two breeds. The progeny known as crossbreds, possess hydbrid vigour, often expressed in females as increased mothering ability. They are sometimes crossed with a male of a third breed to introduce further qualities.

Cross Cultivation. Means the working of land by ploughing the second time at right angles to the first.

Cross-Fruitfulness. Interfruitfulness; the ability of a variety of fruit plant to mature fruit following pollination by another specific variety.

Cross Protection. One strain of a virus which is infecting plant tissues and potect them from infection by other strains of the same virus.

Crossing. Another name of hybridization. It is defined as the artificial cross-pollination between genetically unlike plants.

Crotch. Common joint of large branches from which sometimes more than one lateral branch arises.

Crown. 1. The top part of a plant, such as the top branches of a fruit tree or the leaves of certain root crops, e.g., sugar beet. 2. The junction of the root and stem of a plant. 3. The perennial root-stock of certain plants e.g., rhubarb.

Crown Furrows. Scratch.

Crown Craft. A type of graft in which the scion is fixed into a vertical slit in the bark and sap wood of the stock. Also called a rind graft.

Crucifer. A plant belonging to the cabbage family (*Cruciferae*) characterized by four petals and sepals, both arranged in the form of a cross.

Crushed Grain. Cereal grains crushed in a crushing Mill before feeding to livestock to assist in their digestion. Also called rolled grain.

Crushing Mill. A machine which crushes cereal grains before feeding to livestock, by rolling it between two iron cylinders rotating Close togther. Also called Roller Mill.

Crutch. A pole with a cross-bar at the end, used in dipping sheep to push the shoulders down, immersing the head and shoulders.

Cryogenic Flask. A flask used in Artificial Insemination to transport semen at very low temperatures.

Cub. 1. The young of certain animals e.g., fox. 2. A cattle pen or small enclosure such as a chicken coop or rabbit hutch. 3. A Crib or fodder rack.

Cuber. A machine which forces meal, usually binding it with molasses at the same time, through circular holes to produce extruded pencils of meal which are chopped into cubes or pellets of various lengths ranging from 3.25 mm for poultry to 13 mm for cattle.

Cucumber. A creeping plant cultivated under glass and in the open where there is no frost risk, both as a market garden and vegetable garden crop. Used as a salad ingredient and for pickling.

Cud. The partly digested food brought back from the first stomach or Rumen or a Ruminant to be chewed again.

Cull. An animal separated and removed from the herd of flock, being unsuitable, of poor quality or too old.

Culm. The flowering stem of a grass.

Cultivar. A variety of a cultivated plant.

Cultivation Sheet. The term used for the record which furnishes the complete history of the individual crop grown on the field with its cost of cultivation from preparing the land to harvesting the crop and all other extra charges upto sale, revealing at the end total expenses of the crop, yield, and net profit or net loss.

Cultivator. An implement having number of tynes which are attached to a frame. It is used for tilling the soil between standing rows of crops. It stirs the soil and is able to break the clods. The tynes fitted on the frame of the cultivator have been found to comb the soil deeply in the field.

Cultivator, Tralled Type. Having a main frame which is carrying a number of cross members to which tynes are fitted. At the forward end of the cultivator, there exists a hitch arrangement for hitching purpose. A pair of wheels are fixed in the cultivator. The lift has been operated by both wheels simultaneously so that draft is even and uniform. The height of the hitch has to be adjusted so that main frame becomes horizontal over a range of depth setting.

Culture. Bacteria growing on or in a substances which provides suitable feeding conditions.

Culvert. A channel for carrying water under a road, railway.

Curd. Milk thickened or coagulated by acid (Rennet) in cheese making, and distinguished from the more watery whey.

Cure. The preservation of meat by salting, smoking or pickling for storage.

Curry. To rub down, clean and groom a horse, usually using a special comb called a curry comb.

Curring. A drying process by which most of the moisture in the harvested leaf has been removed. Different methods of curing include air curing, rack curing, smoke curing, pit curing, ground curing, sun curing. It is process by which the harvested produce of tobacco is made ready for the market.

Cusec. Refers to the quantity of water which is flowing at the rate of one cubic foot per second. One cubic foot of water weighs 6.24 lbs i.e., 6.24 gallons. One cusec flowing for one hour will be equal to 62.4 $\times$ 60 $\times$ 60 lbs=101 tons approximately.

Custom. Refers to accepted ways of eating, meeting folks, training the young, supporting the aged, etc.

Custom Inoculation. Refers to the inoculation of farmer's seeds, usually by machine not longer than 3 or 4 weeks in advance of planting.

Cut. 1. A term used to described the steepness of the crest of a Furrow slice. 2. A term used to described the mowing of grassland for silage.

Cutin. Refers to a fatty substance which is formed from a mixture of complex derivatives of fatty acids and impregnated the outer wall of the epidermis of plants and forms a separate layer, the cuticle, outside this wall. It is relatively impermeable to water and gases, restricting water loss.

Cuttage. Plant propagation by cutting.

Cutter Bar. A mechanism on a Binder, Combine Harvester or Mower which cuts the crop and consists of a fixed finger bar (Fingers) and reciprocating knives. The knives of a mower operate at a faster speed than those on binders and combines since grass is more difficult to cut than the dry straw stems of cereals.

Cutter Blower. Forage Blower.

Cutting. A cut section of material removed from a living plant which, when placed in a suitable rooting medium, will produce roots and give rise to a new plant. Cuttings are used extensively in the propagation of certain types of plant, e.g., Chrysanthemums, Geraniums, etc. (Refer to Graft, Budding, Layering).

Cutworm. A caterpillar of various noctuid moths, active at night only, attacking roots and stems and named after its habit of eating completely through young plant stems at ground level. Also called surface caterpillar.

Cycloheximide. An inhibitor of protein biosynthesis.

Cyanamide ($CaCN_2$). Calcium cyanamide, an Artificial Fertilizer, converted by soil water into Ammonia. Also called nitrolime.

Cyst. 1. A membrane secreted around the resting stages of many animals during their development, e.g., that enclosing Eelworm eggs in the soil. 2. A swelling or hollow tumour containing soft or watery matter as distinct from pus.

Cytogenetics. Science dealing with the study of the biological systems using the combined approach of cytology and genetics.

Cytology. Branch of biology dealing with the structure, function, properties, physiology, development and reproduction of cells.

Cytoplasmic Genetic Male Sterility. Type of male sterility which depends upon the action of genes carried in the nucleus with a particular cytoplasm.

Cytoplasmic Inheritance. Means the transmission of hereditary characters via the DNA of self-replicating extranclear organelles like mitochondria and chloroplasts.

Cytoplasmic Male Sterility. Type of male sterility which is governed by the cytoplasm rather than by the action of nucleus. It gets transmitted only through the female parent.

Cytosol. Soluble part of cytoplasm which is left after the particles got separated, e.g., by centrifugation.

Cytotaxonomy. Classification of plants which is based on a study of cell-structure.

D

2,4-D. An abbreviation for 2:4 di-chloro-phenoxy-acetic acid (also abbreviated to DCPA). A translocated herbicide used to control many broad-leaved annual and perennial weeds post-emergence in cereals (except spring oats) and grass seed crops, in grassland and tuif. Closely related chemically to 2,4,5-T.

DDT. Dichloro-diphenyl trichloroethane. A nerve poison which kills insects by contact, but may also act as stomach poison. Its toxicity is due to presence of p-isomer.

D-Horizon. Refers to a sub-surface layer or rock stratum which is neither parent material nor the source of parent material for the soil.

D Trials. Series of large plot trials which are carried out for comparison of varieties using farm methods of seeding and harvesting.

D-value. The percentage of digestible organic matter in the dry matter of animal feeds such as hay, silage and dried grass.

Dairy. 1. A farm building, often a single room, in which milk is cooled and temporarily stored prior to collection for transportation to a commercial dairy (see 2 below), and in which it may be treated and made into cream, butter and cheese—a practice now becoming rare on farms. Milking equipment is usually washed and kept in the dairy. 2. A commercial processing plant to which fresh farm milk is transported by tanker (Bulk Milk) for treatment and distribution as bottled milk, and for the production of cream, butter, cheese, yogurt and other products for retailing. Some dairies are purely milk bottling plants whilst others are mainly creamers.

Dairy Cows. Cows and heifers kept for producing milk or for rearing calves for a dairy herd, as distinct from Beef cows.

Dairy Farm. A Farm on which the principal activity is milk production.

Dairy Husbandry. Deals with the care, breeding, feeding, and milking of dairy cattle, and the production and sale of milk.

Dam. The mother of an animal. Usually used when describing the pedigree of a farm animal. (Sire)

Dumping-off. A disease of seedings caused by *Pythium* and various other fungi in excessively moist conditions. It causes wilting and roots to die.

Dandruff. Accumulation of brainlike scales on the skin, becomes of exfoliation of the sebneceous glands.

Dark Seed. Seed which will germinate, only when kept in the dark, when other conditions would normally favour germination.

Day-old-chick. One that is depatched to the buyer within 24 hours of hatching.

Dead-in-shell. In general terms a fertile egg in which the developing embryo has died. In specific terms, embryo death occurring between the 14th and last day of incubation.

De Man's Values. Refer to the measurements and ratios to express length of nematode, its relative width, relative lengths of esophagus and tail, and position of vulva.

De-Suckering (Tobacco). Removal of the suckers is known as desuckering.

Dead Furrow. Refers to the open trench which is left in between two adjacent strips of land after finishing the plough.

Dead Seed. Seeds which are not capable of germinating.

Deadstock. The equipment required to operate a farming enterprise, such as tractors, implements, hay racks, and hurdles etc.

Deadweight. The weight of a dressed carcase.

Dead Water. Standing or still water.

Deassimilation. Refers to the utilization of food by a plant.

Debearder. Preconditioning machine which is used in seed testing laboratory to debeard barley, breakup lucerne seed pods, breakup grass seed doubles, separate and polish vegetable seeds, decorticate sugar beet seeds and remove extra glumes and hulls.

Decalcification. 1. Means the removal of calcium carbonate or calcium ions from soil by leaching, a natural process in soil formation. 2. Also refers to the removal of calcium from the bones of animal in negative calcium balance.

Decarboxylase. An enzyme which is able to liberate carbon dioxide from the carboxyl group of a molecule.

Deciduous. A descriptive term of plants, particularly trees, which shed their leaves annually, usually in the autumn. (Evergreen)

Decidous Teeth. The first teeth to appear in a young animal, later shed and replaced by permanent teeth. Also called milk teeth.

Decortication. The removal of husks from seeds. Oil seeds are frequently decorticated before oil extraction. Cattle Cake derived from oil seeds, (*e.g.*, Cottonseed Cake) may be either decorticated or undecorticated (the husks having been left on).

Deduplication. Chorisis, *i.e.*, increase in parts of floral whorl which occurs due to division of its primary members.

Deep Litter. A system of Bedding for cattle or poultry using straw shavings or sawdust.

Deep Percolation. Refers to the moisture which penetrates below the depths from which it may be used by plants. It represents that part of the water absorbed which exceeds the field capacity of the soil within the depth of root development.

Deep Percolation Loss. Water percolating downwards through the soil beyond the reach of plant roots.

Deferred Grazing. Means withholding animals from a pasture beyond the normal beginning of the grazing season.

Deficiency Deseases. Diseases of animals and plants due to an inadequacy of one or more essential food substances, such as a Vitamin, mineral Element, Protein of Amino Acid, etc., *e.g.*, Grey Leaf, Brown Heart.

Deflocculation. The dissociation of large particles into smaller ones. Particularly applied to soils and the disintegration of crumbs under wet conditions.

Defoliation. The removal of the leaves of a plant. A defoliant is a chemical which causes this.

Deforestation. The process of clearing forests.

Degras. A fat derived from sheepkins.

Degree Brix. Measure for the sugar content of a solution.

Dehiscent. A descriptive term for fruits which spilt open to release the seeds, *e.g.*, peas.

Dehorning. The removal of an animal's horns to prevent it wounding others. The best method is to disbud the young animal, but horns can also be removed under anaesthetic by electric saw. Also called dishorning.

Dehulling. Refers to the removal of the outer covering from intact grains of seeds.

Dehydrated Foods. These are the products from which most of the water has been removed in order to improve their stability during storage.

Demography. Refers to the study of numbers of organisms in a population and their variation with time.

Demulcent. Drug having a soothing action especially upon the mucous membrane or inflamed skin surface.

Denaturing. The treatment of wheat grain so as to make it unusable for human consumption either by staining with a suitable dye or by the addition of fish oil to give a smell to it. Wheat, when denatured, can be used in animal feed, but not for flour-milling.

Dendrochronology. Means the dating of trees by observation of annual growth rings in the trunks.

Dendroclimatology. Determination of past climates by studying of relative widths of annual rings.

Denitrification. The breakdown of nitrates and nitrites by denitrifying bacteria in the soil in anaerobic conditions with the release of gaseous nitrogen.

Densitometer. Instrument for measuring photographic density.

Density Crop. Compactness of tree stocking which is expressed as a decimal co-efficient, by taking normal numbers, basal area or volume as unity and assessed by comparison with yield table figures.

Dent Corn. Horticulturally produced variety or corn which is characterized by a depression in the end of the dry kernel.

Dental Formula. A formula indicating the number of each kind of teeth for a given animal species. The teeth are shown in the order incisors (i), canines (c), premolars (p) and molars (m), with teeth numbers for one side of the upper jaw shown above those for the lower jaw. The formula for a pig for example is

$$i\,\frac{3}{3},\ c\,\frac{1}{1},\ p\,\frac{4}{4},\ m\,\frac{3}{3}.$$

Dentition. The cutting or growing of teeth. Also the arrangement and number of teeth.

Depreciation. A figure included in the annual farm accounts as an expense indicating the amount by which various assets such as machinery and implements, etc, have lost value with ageing.

Depth Wheel. A wheel on an implement used to control the depth of working.

Dermatitis. An inflammation of the skin.

Dermatology. Refers to the scientific study of the structure and function of skin, diseases of skin, and the treatment of skin diseases.

Derris. An insecticide powder from the root of the tropical plant, *Derris elliptica.* Used against fleas, lice and Warbles.

Designated Diseases. Plant diseases which have been specified for seed certification of a given kind and in those regard certification standards must be met.

Designated Weeds. Species of weeds have been specified for seed certification of a given kind and in those regard certification standards must be met.

Dessicant. A chemical applied to a crop causing the foliage to dry up so that harvesting can commence.

Detassel. Means the removal of young tassels in maize plants so as to render the plants fully female, specially in hybrid maize seed production.

Detoxicants. Agents which are used to neutralize the action of poisonous substances.

Detritus Food Chain. Food chain in which microorganisms are able to absorb and break down the energy-rich compounds synthesized by the primary producers.

Dew-claw. The rudimentary fifth digit protruding from the heel of cattle, pigs and dogs.

Dewlap. 1. The losse flesh which hangs from the throat of oxen and cows. 2. The fleshy wattle of the turkey.

Dextrins. Refer to the products which are obtained by the partial chemical breakdown of starch by the action of heat, acids, enzymes, or a combination of these agents.

Dhaincha (Ind.). *Sesbunia cannabina* (Retz.) pers. (*S. aculcata* Pers var. *cannabina* Baker). Family Papionaceae. Ranks next to sannhemp as green manure crop; a legume which has been adapted to a wide range of soil and climatic conditions, growing well on soils too wet, too dry, or too salty for most green manure crops. It is grown as green manure mostly in low lands, and planted on paddy bunds or in high lands for seed production.

Diallel Crossing. Means type of crossing in which each of a number of males is made to cross to each of a number of females so as to compare the breeding value of males.

Diamond Tup. A shearling male lamb. Also called a dinmont, particularly in the borders.

Diaphragm Pump. A type of pump in which a flexible diaphragm, generally of rubber is the operating part.

Diarrhoea. Waterly faeces which are caused by internal parasites, bacterial infection, excessive feeding of green succulent or other feed. Simple diarrhoea could be treated with catharitics like castor oil.

Diastase. An enzyme produced in germinating seeds and by the pancreas which converts starch to sugar.

Dib. A small bone in a sheep's leg.

Dibber, Dibble. A pointed hand tool used to make holes in the ground for seeds or plants.

Dibble, To. To sow seeds, or plant seedlings, in holes made with a planting peg.

Dibbling. Refers to the method of sowing the crops by using manual labour or a dibbler where in specific spacing and number of plants are maintained between the rows and within the rows.

2, 4-Dichlorophenoxyacetic Acid (2, 4-D). A synthetic auxin which is used as an herbicide and defoliant.

Dieldrin. An insecticide ten times more powerful than D.D.T. A Chlorinated Hydrocarbon, highly poisonous to birds and fish, banned from use in sheep Dips in 1966 and as a winter wheat dressing in 1975. Its use is now limited under the Pesticides Safety Precautions Scheme to specified circumstances where no suitable alternative exists.

Diet. Food or feed of an animal, which is having implications of amounts consumed, or offered, in any given time.

Dietary Diseases. Diseases caused if the animal needs for nutrients and vitamins are not satisfied.

Dietetics. Deals with the application of the science of nutrition to the feeding of individuals and group of people.

Digestibility (Indicator Method). A quick method which is used for determining the digestibility of animal feed without measuring either the feed intake or faces output. This is carried out by using an indicator with the feed to be tested and then finding out the ratio of the concentration of the indiator to that of a given nutrient in the feed and also in the faces resulting from feed, as shown below:

Digestibility =

$$100 - \left[100 \times \frac{\%\ \text{indicatorinfeed}}{\%\ \text{indicator in faeces}} \times \frac{\%\ \text{nutrient in faeces}}{\%\ \text{nutrient in feed}}\right]$$

Digestible Crude Protein. A measure of the protein value of a Feedingstuff, determined by a Digestibility Trial. It is more useful than Crude Protein or True Protein content in formulating the diets of Ruminants since they can vary considerably in digestibility and composition.

Digestible Energy Value. The amount of utilisable energy in an animal food after deduction of the energy value of undigested faecal residues. (Digestibility Trial)

Digestion Coefficient. The percentage digestibility of a constituent of a foodstuff (e.g., Crude Fibre, Protein) calculated from the results of a Digestibility Trial.

Digestive Juices. Various juices secreted into the digestive tract of animals which cause chemical changes to food and facilitate its absorption into the body, e.g., Saliva, Bile, Gastric Juice, Pancreatic Juice and Intestinal Juice.

Digger. A type of mouldboard, abruptly concave and twisted, that produces a rough well-broken furrow slice, as distinct from the continuous smooth slice of the Lea Plough. Used for deep ploughing and popular for general work.

Dioecious. A term applied to plants having the male and female flowers borne on different individual plants of the species e.g., willow.

Dip. 1. A proprietary chemical which is diluted in a recommended volume of water in a Dipping Bath for Dipping animals, mainly sheep. Dips containing Dieldrin were more advantageous than arsenic and sulphur based dips as they persisted in the fleece. They significantly reduced strike in sheep. Dieldrin was banned from use in sheep dips in 1966, being highly poisonous, and a possible meat contaminant. Nowadays dips mainly contain organo-phosphorus chemicals. 2. A Dipping Bath.

Dip Slope. The long gentle slope of a line of hills following the inclination of the rock strata, as distinct from the shorter, steeper Scarp Slope.

Diploid. A term applied to living cells having two sets of chromosomes in the nucleus, present in pairs. Characteristic of most animal and plant cells. In reproduction, the pairs separate in the process of meiosis producing gametes which are haploid, having only one set of chromosomes.

Dipped Sheep Market. A market, or a separate and distinct part of a market, at which, for a particular period, only dipped sheep are sold.

Dipping. Consists in immersing animals in an insecticidal liquid.

Dirt Eating. The term used for the habit of young animals which sometimes eat litter, soil, wool and other indigestible substances that may lead to the formation of balls of foreign material in the stomach, which may lead to death due to stoppage of the exit from the stomach.

Direct Drilling. The sowing, by drill, of seeds direct into a field, without previous cultivation and usually following the application of weed control sprays. The lack of tillage reduces the effectiveness of soil-acting herbicides and encourages certain weeds in winter-sown crops, and intermittent ploughing is now advised.

Direct Reseeding. Sowing grass seeds (usually a Seeds Mixture) without a Cover Crop in a field in which the immediate past crop was grass.

Direct Seeding, Direct Sowing. Sowing grass seed (usually a Seeds Mixture) on bare ground without a Cover Crop.

Dirty Land. Land full of weeds.

Disbud. 1. To remove buds from a plant. 2. To remove or prevent the growth of the horn buds of calves by painting them with a caustic compound or by use of a hot iron, usually in the first 7-10 days after birth. Kids and lambs are also sometimes disbudded.

Disc Colour. A flat circular kind of colour which rotates as the plough moves and makes a vertical cut in the soil, separating a slice of soil from unploughed land, which the share and mouldboard then turn to produce a clean furrow slice.

Disc Harrow, Disc Cultivator. A type of harrow with a number (normally four) of sets or gangs of sharp-edged concave rotating discs, the angle of which can be varied in relation to the direction of travel, altering the severity of cultivation according to requirements. The gangs can be positioned so that disc faces in different gangs face in opposite directions. Often used on light land, and on freshly ploughed grassland or stubble since unrotted trash is not brought to the surface.

Disc Plough. A type of plough with heavy dish-faced wheels, individually mounted, and usually adjusted at an angle (between 35 and 45°) to the line of travel, which cut and turn the soil to the side, partly inverting it. The discs are usually between 50-80 cm (20-32 in.) in diameter and are sometimes fitted with scrapping devices to remove accumulating mud. The discs may be plain or cut-away. The latter give more bite and cut into hard soils. Disc ploughs are used for deep, rapid cultivations on rough land but are not very common Mould board ploughs produce better quality work.

Disease. Any disturbance which interferes with normal structure, function or economic value of host.

Disease Cycle. Chain of events which are taking place in the development of a disease.

Disease Endurance. Refers to the ability of plants to tolerate the invasion of a pathogen without showing much damage and many symptoms. Synonym, disease tolerance.

Disease Escape. Refers to the ability of susceptible plants to avoid disease attack due to some inherent qualities such as rapid growth and early maturity or environmental conditions like changes in sowing time, growing period and method of cultivation. Synonym, klendusity.

Disease Resistance. Refers to the ability of plants to withstand, oppose, or overcome the attack of pathogens.

Disinfection. 1. The cleansing of appropriate buildings, animals, implements and utensils, etc., which may be hardbouring harmful bacteria or viruses, after an infection has existed on a farm. The process can involve physical cleaning, the application of approved chemical solutions, steam cleaning and fumigation. 2. The cleaning of farm equipment and utensils (e.g., milking plant) using suitable detergents and disinfecting agents (e.g., sodium hypochloride) to reduce bacterial contamination. 3. The chemical treatment of seed before sowing to control fungal disease in crops, particularly cereals.

Disinfestant. Substance which is able to kill or inactivate pathogenic microorganisms in the environment or on the surface of plant parts before infection.

Disinfestation. External parasites Mechanical or physical means of hindering their development *e.g.,* cleaning the yards of all refuse, removing litter and droppings frequently etc.

Dispersed Soil. A type of soil in which the clay is readily to form a colloidal soil, usually with a low permeability, which upon drying shrinks, cracks, and becomes hard, and upon wetting slakes readily and is plastic.

Dispersing Agent. Substance added in water dispersible insecticide powders and concentrate pastes to obtain uniform concentration of spray all over treated area. Common examples include protective colloids such as sodium carboxy and methyl celluloses.

Distillers' Grains. A by-product of whisky manufacturing similar to Brewers' Grains consisting of the remains of malted barley. Used as an animal Feedingstuff either wet or dry, but mainly preferred dry as wet grain retains raw alcohol which may intoxicate the animals.

Distillery Waste. The form used for the residue which is obtained from the fermentation and distillation of beer and distilled spirit. It contains considerable amount of potash and some ammonia recovered and used as fertilizer material.

Ditch. A trench dug in the ground to drain water from farmland, usually discharging to a stream or other watercourse. The water reaching a ditch may be by surface run-off, by natural underground flow, or via an artificial underfield drainage system, the drains of which discharge into the ditch.

Ditcher. A machine designed to excavate, clean or repair a ditch.

Ditch Water. Stagnant, foul-smelling water such as that often found in ditches.

Diviner. One who divines or detects the presence of water beneath the ground (Dowser)

D.N.O.C. Di-nitro-ortho-cresol. An insecticide, acaricide, and powerful weedkiller applied as a foliar spray.

Dock. 1. A weed (*Rumex sp.*) with large leaves and a long root. 2. To remove the whole or part of an animal's tail. Low-land breeds of sheep are customarily docked as lambs to avoid the accumulation of dirt and faeces in the tail.

Doddle. Loose flesh hanging from the throat of a goat. Also called tassel or wattle.

Doe. A female deer, hare or rabbit.

Dolly. An ornament made of straw usually placed on top of a stack. Also called corn dolly.

Dominant Plant. The major plant in a vegetation community, usually the most numerous or the tallest, e.g., beech wood.

Dorking. A table breed of fowl, square-bodied, variously coloured (including red, white, silver-grey and barred), with a single comb and five-clawed white legs. Hens produce white eggs.

Dormancy. A state of low metabolic activity in living organisms, particularly in seeds and spores, usually associated with unfavourable environmental conditions (e.g., low temperature, inadequate moisture, short day length) for growth and thought to be controlled by hormones. Stored seeds can remain dormant for years and will germinate after sowing if conditions are right. Deciduous trees lie dormant in the winter after shedding their leaves.

Dormant Spray. A spray applied to a crop during a dormant period.

Dormant. Resting condition for a living plant or animal with a relatively inactive metabolism.

Dormant Bud. A bud remaining inactive for an indefinite period, until stimulated into growth.

Dormant Spore. The spore which fails to germinate under the same nutritive and environmental conditions which later allow the formation of germ tube.

Dorset Wedge Silage. Silage produced by filling a silo in wedge shaped layers with cut grass or other greenstuff, covering with plastic sheeting and sealing the silo to ferment.

Double Chop Harvester. A type of offset-trailed or semi-mounted Forage Harvester, consisting of a cutting or pick-up mechanism of the flail-type, with an auger delivering the cut crop sideways to a flywheel chopper mechanism for further chopping before it is loaded by chute into the trailer. This type of machine produces a non-uniform chop length. Also called a Flywheel Chopper.

Double-Dropping. Means growing a second crop in one growing season after the first crop is harvested from the same piece of land (overlap between crops can occur).

Double Mounting. The fitting of dual wheels to the rear axle of tractor in an effort to reduce soil compaction by increasing the contact area, and to increase traction.

Doubles. Twins, especially lambs.

Double-suckling. A method of feeding beef cattle in which a second calf is introduced and allowed to suckle alongside a cow's own newborn calf. The second calf' is sometimes rubbed with part of the Placenta to encourage the cow to accept it as her own. The system is only used with cows producing sufficient milk to feed two calves.

Double-work. A method of ensuring a successful fruit-tree Graft between a Scion and Stock which otherwise would be incompatible, by grafting between the two a scion of a variety compatible with both, called an intermediate stem piece.

Down. The soft plumage covering young birds, or that found under the feathers of certain birds (*e.g.*, ducks, geese).

Down Breeds. Breeds of hornless, short-wooled sheep, with coloured faces and legs, producing good dense fleeces and high quality meat.

Down-Calver. A cow or heifer about ready to give birth to a calf.

Dowse. Chaff.

Dowser. A water divine—a person able to locate underground water using a dowsing-rod (a forked twig) which quivers when over a water source.

Draft. The force exerted on a trailer or other trailed implement by a tractor, measured in newtons (N). Also called draught and drawbar pull. (Dynamometer)

Draft of Mouldbord Plough. According to Collins, the draft of a plough on the ground has been 18%, draft due to turning furrow slice has been 34%, draft due to cutting slice is 48%. Thus it is seen that practically 50% of the total draft of the plough has been used in cutting the furrow slice.

$$HP = \frac{\text{Force (kg)} \times \text{distance travelled (ft / mt)}}{33000}$$

in FPS unit

or

$$HP = \frac{\text{Force (kg)} \times \text{distance travelled (metre / mt)}}{45000}$$

in metric unit.

Drafted Animal. An animal either added to (drafted in) or removed from (drafted out) a herd or flock. Draft ewes are those sold from a breeding flock of sheep whilst they still are able to produce lambs. Usually they are sold from a hill flock and sent to a lowland farm when they are considered to lack hardiness for continued hill breeding, normally after producing three crops of lambs.

Drag. A kind of fork with curved prongs at right-angles to the handle, used to draw manure off a load or from a heap.

Drag Harrow. A kind of heavy harrow, with tines having curved ends, and often with wheels, capable of quite deep work. The action is similar to that of a cultivator which has fewer tines and can work deeper. Drag harrows are preferred on difficult land for seed bed preparation, bringing fewer unweathered clods to the surface. Also called a duck-food harrow.

Drail. The iron bow of a horse-drawn plough from which the traces draw.

Drain. 1. An artificially dug channel or ditch to carry surplus water away from farmland. Common in the Fens. Usually maintained by an Internal Drainage Board. 2. An underground channel in a field drainage system usually comprising a piped duct of clay, porous concrete or plastic tube.

Drain Gauge. A tank placed in a ditch beneath a land under-drain outlet to catch and measure the volume of water passing through the drain.

Drained Honey. Honey obtained by draining uncappped broodless honeycombs.

Draining Pen. A pen into which sheep emerging from a Dipping Bath are directed to allow surplus dip to drain off the fleece and return via a sieving device to the bath.

Drain Jetting. The practice of flushing drainage systems with water in areas where sedimentation and ochre build-up occurs in the pipes.

Draught Horse. A horse used to pull carts, trailers and farm implements (*e.g.,* the Shire horse).

Drawbar. A metal bar at the back of a tractor to which implements are attached.

Drawbar Horsepower (DBHP). Refers to the power of a tractor measured at the end of the drawbar on the hitch point. It is used to pull loads.

Drawbar Power. Power of the tractor, which is measured at the end of the drawbar. It is the power available, to pull loads or draw machines.

Drawbar Pull. Pull exerted by a tractor at its drawbar end.

Dray. A low strong cart with no fixed sides, used for carrying heavy loads.

Dredge Corn. A mixture of cereals sown together (*e.g.,* oats and barley) producing a yield of grain better than if either crop were grown separately. (Mashlum).

Drench. A liquid medicine poured down an animal's throat.

Dress. In general terms to clean prepare or repair. 1. To apply fertilizer to land (*e.g.,* Top Dressing.) 2. To chemically treat seeds before sowing to control fungal disease in growing crops, particularly cereals. 3. To clean corn seeds by removing Chaff, weeds and other unwanted matter. 4. To remove the tough fibres from flax. 5. To sort potatoes into various sizes. 6. To prune a hop plant by cutting the previous year's growth from the crown. 7. To recut grooves in a millstone.

Dressed Carcase. Carcase.

Dried Blood. An Organic Fertilizer derived from waste blood from slaughterhouses, dried by evaporation. It has a nitrogen content of about 13%.

Dried Grass. Grass which, after cutting, has been thoroughly dried by artificial means, and is used as an ingredient in animal feedingstuffs. It is either milled and mixed with molasses to produce cubes or pellets, or pressed directly into cobs or waters. Used for many years for poultry and pigs, and now becoming poupular as a dairy cow ration combined with a concentrate and roughage. Sometimes called dehydrated foddeer.

Drill. 1. An implement used to sow seed in rows (Seed Drill). Seed and fertilizer may also be sown simultaneously using a Combine Drill. 2. A ridge with seed or growing plants on it (*e.g.*, turnips, potatoes), or the plants in such a row.

Drill Barrow. A hand drive Seed Drill.

Drill Harrow. 1. A light flexible Chain Harrow trailed behind a Seed Drill to cover the seed with extra soil. 2. A drill mounted on a Spring Tined Harrow. Also called a tine harrow.

Drill Plough. An old type of trailed plough with an apparatus attached to the beam which drilled seed directly into the newly produced furrow.

Drilled to a Stand. Seeds sown by a Seed Drilll at predetermined intervals so as to produce a crop with a specific plant population which requires no thinning. (Thin).

Drone. A male bee.

Dropsy. Means the excessives accumulation of waterly fluid in any of the tissues or cavities of animal body.

Drought. An extended period without rain.

Drought (Agriculture). Shortage of moisture for crop growth or animal production.

Drought (Resistance). (1) Characteristics of plants which are suitable for cultivation in dry condition regardless of the inherent mechanism that provides resistance. One of the more important properties is the capacity to endure, without injury, an intense loss of water. (2) Refers to the relative ability to maintain growth yield under moisturestress conditions.

Drought Year. When rain fall is short by more than twice the deviation, year is regarded as a drought year for a particular place.

Drove. 1. A number of cattle or sheep being driven from place to place. 2. A track along which cattle or sheep are driven. Also called drift, drift-way.

Drum. The cylinder of a Threshing Machine or Combine Harvester bearing Beaters, which revolves and dislodges the grain from the ears of cereals. Also called a threshing drum.

Drumhead. A kind of Cabbage having a flattish round head.

Drup. A kind of fruit comprising a hard kernel containing a single seed enclosed within a fleshy pulp, *e.g.*, plum, cherry. Also called a stone-fruit.

Dry Bible. A disease of horned cattle in which the third stomach, or bible is very dry.

Dry Cow. A cow not producing milk.

Dry Cured Bacon. Bacon preserved by being rubbed with dry salt as opposed to being soaked in brine.

Dry Farming. Refers to the type of agricultural operation which involves deep cultivation of the soil to form a sufficient reservoir for the moisture as it falls, surface cultivation to prevent or reduce evaporation; selection of drought resistant crops.

Dry Feeding. A method of feeding meal to animals in a dry state without the addition of water. Used with pigs and poultry.

Dry Mash. A mixture of feedingstuffs fed to animals in a dry state.

Dry Matter. The various mineral and organic components (*e.g.* Ash, Crude Protein, etc.) in feedingstuff derived by Proximate Analysis.

Dry Milking. Also called dry hand milking. Complete milking to ensure that the udder of cow has been completely empty or dry.

Dry Measure. A system of measuring by bulk. (Bushel, Peck).

Dry Period. A period between Lactations when a cow is allowed to rest from the strain of producing milk, usually of 6-8 weeks duration.

Dry Planting (Sugarcane). Refers to the practice of planting the cane in dry condition of soil and irrigating later on. This method of followed in heavy black soil.

Dry Rot. 1. The decay of timber caused by fungi. The wood is destroyed and reduced to a dry, brittle, powdery mass. 2. A storage disease of potatoes caused mainly by the fungus. Spores infect the tubers via woulds and bruises caused by rough handling during harvesting. Tubers are caused to wither and wrinkle.

Dry Roughase. Bulky; dry, non-succulent animal foods (*e.g.*, hay and straw), rich in Crude Fibre, used mainly in Maintenance Rations for Cattle and sheep. Also called coarse fodder.

Drying-off. Gradual reduction of the volume of milk taken from a cow so as to cause it to cease lactating.

Drying front. Refers to the advancing zone of the bin of drying grain which includes all the grain which has just reached some arbitrarily initial moisture content, the zone between the initial and final moisture content.

Dual-purpose Breeds. Breeds of (*a*) cattle considered useful for both beef and milk production, (*b*) poultry considered good both for egg laying and as table birds, and, (*c*) pigs, the females of which are kept generally for crossing with either bacon or pork type boars.

Dubbing. The removal of the comb of a cock. Also the trimming of part of the comb in Day-Old-Chicks. The practice reduces frost bite, and is advantageous of serious Cannibalism or pecking occurs.

Duck. A smaller bird of the duck family (*Anatidae*). The female adult duck as distinct from the male adult duck or drake.

Duckling. A young duck, usually less than 12 weeks old.

Duff. The organic layer in forests. It consists of fallen vegetative matter in the process of decomposition, including everything from pure humus below to the litter on the surface. Duff is a general, non specific term.

Dug. A niple or udder, particularly of a cow.

Dump Box. A large portable container similar to a Forage Wagon, placed adjacent to a Silo, into which trailer loads of Forage from the field are tipped, and from which the forage is delivered in an even steady flow, usually by a moving floor mechanism or elevator belt, for loading into the silo. It may also be used to deliver forage directly into a Manger.

Dump Rake. An expression for hay rake.

Dung. The faeces of animals. Also to excrete in while grazing, or to spread it on the land.

Dung Liquor. The dark coloured liquid which drains from a manure heap, containing soluble compounds of ammonia and organic matter.

Duster. Machine to apply chemicals in dust form. Dusters make use of airstreams to carry pesticides in finely divided dry form on the plants.

Duster, Pluger Type. Simple duster having a small piston. The piston drives a current of air over the dust in the hopper.

Duster, Power Operated. It mainly consists of a power driven fan, a hopper and a delivery spout. The fan creates strong air flow which makes the dust to blow off from the hopper to a considerable distance either vertically or horizontally.

Duster, Rotary Type. Duster with a hand operated rotor in front of the operator. For dusting tall crops, more force of delivery is required and hence rotary dusters are preferaed.

Dusting. The application of insecticide or fungicide to crops in a dry, powdery state. It is less efficient than spraying, usually requiring about twice the amount of chemical per unit area.

Dwarf Trees. Trees made smaller than the size commonly accepted as normal for the species and variety under consideration.

Dye Reduction Test. Test of measurement of activity of bacteria in a sample of milk. It is the measurement of rate of consumption of oxygen in milk due to bacterial growth and activity. Indicators such as methylene blue or resazurin are used which decolorize milk when oxygen level in its falls. Time taken for decolorization provides indication of bacterial numbers and keeping quality of milk.

Dyke. 1. A ditch or minor lowland watercourse. 2. A raised bank of earth, stones, etc., constructed to keep a river or sea water from flooding adjoining, usually low-lying, land. Also called Dike.

Dynamometer. An instrument which measures power. Used in agriculture to measure the resistance or Draft of any implement drawn by a tractor.

E

Ea. A term used in the Fens for a drainage, canal, probably derived from the French word *eau,* meaning water.

Ean. To give birth to a lamb. Also called yean.

Eanling. Ayoung lamb.

Ear. 1. The spike or flowering head of grass, usually applied to cereals. Also to form such ears. 2. An obsolete term meaning to plough or till.

Early Bite. A sudden growth of grass which can be grazed before the main crops of grass has really begun to grow.

Earmarking. The marking of animals ear for identification, by attaching a marked metal or plastic tag, or cutting distinctive notches.

Earth. 1. A fox's burrow. 2. To cover with soil.

Earth-board. A Mouldboard.

Earth-Nut. Another name for round nut.

Earthing up. The building up of a ridge with soil removed from adjacent furrows, particularly used to cover tomatoes during planting.

Ease. To milk a cow a little in order to relieve pressure on its udder.

Easy Feed. A method of livestock feeding in which the stock are allowed easy access to feed. Usually the feed is transferred mechanically to large hoppers or feeding passages.

Eatage. A obsolete term for grass used for grazing as distinct from for hay, particularly aftermath. Also the right to use aftermath for grazing.

Ecdysone. Hormone which is inducing moulting in insects.

Ecology. The study of the relationship between living organisms and their environment

Economic Botany. Branch of botany dealing with the study of various uses of plants. It includes methods for their better utilization and improvement by mankind.

Economic Irrigation Efficiency. Means the ratio of the actual income attained with the operating irrigation system, compared with the income expected under ideal conditions. It is relating the final output to input.

Economic Protection. Protection of crops from pest damage only to the extent that the resultant economic gains exceed the costs of protection.

Ecosystem. Means complete ecological system of an area, including the plant, animal and other environmental factors.

Ectocommensal. Refers to an organism that lives on the external surface of another organism, the host, without either benefiting or injuring it.

Ectoparasite. A parasite living on the outside of its host, e.g., a Louse. Also called exoparasite.

Ecotophyte. External plant parasite, that is one that lives outside its host.

Ectozoon. External animal parasite.

Eczema. Inflammatory condition of the skin which is not a contageous disease but is usually associated with digestive dearangement, improper feeding, and insanitory environment. Affected animal suffers from an etching of the ears or leggs and soon raw, bleeding areas appear. Treatment involves washing the affected area with a failry strong saponated cresol solution, permitting it to dry and then applying an antiseptic dusting powder.

Edaphic. Of or pertaining to the soil.

Edaphology. The study of the influence of soils on living organism, particularly plants, including man's use of land for growing crops.

Effected Field Capacity (Plough). Actual area which is covered by plough and based on its total time consumed and its width.

Effective Ground-Water Velocity. Actual or field velocity of ground water which is percolating through waterbearing material. It is a measured by the volume of ground water passing through a unit cross-sectional area divided by the effective porrosity. It is also called actual velocity : field velocity.

Effective Rainfall. Precipitation which falls during the growing period of the crop and is available to meet the evapotranspiration requirements of crops. It does not include precipitation lost through deep percolation below the root zone or through surface runoff.

Effluent. Liquid, solid or gaseous water material, e.g., Slurry, industrial waste sewerage.

Egg. 1. An oval-shaped body laid by female birds (and some other animals) comprising a calcareous shell containing Albumen and a Yolk, the latter developing to produce a new young birds which hatches as a Chick. 2. An ovum of femal Gamete. Also called egg cell.

Egg Bound. A term for a bird unable to lay an egg.

Egg-Case. Protective covering or case for transport of eggs.

Egg Classes. There are three quality classes under current E.E.C. marketing regulations : Class A ('fresh eggs'); Class B ('second quality or preserved eggs); and Class C ('non-graded eggs intended for the manufacture of foodstuffs for human consumption').

Egg Tooth. A horny knob on the bill of a newborn birds used to crake the eggshell in hatching, and discarded soon after.

Egg Weight Grades. There are seven weight grades under E.E.C. regulations for Class A and B eggs. Grade 1 are 70 g or over, Grade 7 are under 45 g.

Egg Yolk. Central yellow portion of a hen's egg which contain the egg cell.

Egg White. Albumen.

Eggery. A place where eggs are laid.

Eggler. A dealer in eggs.

Elaiosome. An outgrowth from the sufrace of a seed, having fats or oils. They are sometimes attractive to ants, hence aiding in seed-dispersal.

Electric Dog. A machanism to encourage cows to enter a Milking Parlour from a collecting yard, comprising an electric wire sandwiched between two parallel side wires which is drawn up behind the cows gradually as they go through the parlour, by means of a cord in the parlour. (Electric Fence)

Electric Fence. A system of mobile fencing usually used to facilitate Controlled Grazing, consisting of easily moved insulated posts, usually about 3 ft high, supporting thin wires (usually one for cattle and horses, two for sheep and pigs) carrying an electric current, either battery generated or from the mains. The current is not continuous but in 'pulses' at the rate of about 60 per minute.

Electronic Colour Sorter. A machine which finds use in seed testing laboratory for separating seeds on the basis of difference in colour. Each seed is viewed by photoelectric cells in comparision with a selected background.

Electrostatic Seed Separator. A machine which is used for separating the seeds on the basis of their differential electrical properties. The seed are separated by allowing them to fall free in space, and then using an electric field for deflecting some seeds from their normal flight path.

Elevator. A machine for raising a cut crop, Grain, or Bales to the top of a Silo, storage bin, or bale stack, etc., usually for storage. Various designs, exist. Grain elevators usually consist of a continuously moving belt bearing buckets, or an endless chain bearing flights or steps (operating like a moving staircase). Elevators for hay or straw bales often comprise a Continuous moving chain bearing spikes or cross slats.

Elevator Digger. An implement for harvesting potatoes. There are three kinds; trailed, semi-mounted, and fully mounted. They lift potatoes from the field rows and separate the soil from the crop on a chain web and deposit them in a row on the field for collection. The Potato Harvester is essentially an improved elevator digger bearing a platform for the pickers to stand on, and a system of elevators delivering the potatoes to a sorting table, and conveying them to the side for collection and transport. Some complex modern machines incorporate electronic automatic sorting and rquire only one operator.

Elite Plants. Plants which in trials and tests have been proved to be genetically superior to other plants of the same species.

Elite Variety. An improved variety which is developed by plant breeders and released to farmers due to its superiority in at least one respect.

Elutriation Method. Type of mechanical analysis of soil where the separation of different soil fractions is carried out by washing out particles in a rising current of water (particles whose settling velocity is equal to or less than the velocity of the rising water will be taken away by the current).

Eluviation. Process of removal of organic material in solution or in suspension from the soil by percolating waters.

Embden. A heavy white breed of goose, with blue eyes.

Emblements. Crops raised by cultivation (e.g., wheat, potatoes) as distinct from the fruits of trees or grass.

Embryo. A young animal or plant in its earliest stages of development after fertilization, e.g., a young animal in its mother's womb, or the innermost part of a plant seed.

Embryo Transfer. A procedure which involves the removal of a developing Embryo from a mother's uterus at a very early stage and transfer into the uterus of a recipient female. Foetus development occurs inside the surrogate mother which serves as an incubator and source of milk until weaning, but has no genetic influence on the foetus. The natural mother is left free to produce more fertilised ova to be carried to term by other recipient females or by herself. Embryo transfer (ET) is now commercially applied to cattle, sheep goats and pigs, but was first practised on rabbits in 1890. Freezing techniques now allow the preservation and transportation of embryos in banks and containers, just as with bull semen.

The use of inferior females as recipients can allow them to produce superior offspring every year. The technique of supervaluation is now often combined with ET. This involves treating cows with follicle stimulating hormone (FSH) which commonly causes the release of 8-10 ova in one flush which can be removed and implanted in recipients.

The advantages of ET are that the progeny of a superior female can be multiplied many times, large numbers of genetically similar progency of the same age for reproductive or testing purposes can be obtained, superior progeny can be readily transported in frozen embryo stage, extra progeny from old but valuable females can be recovered, rare breeds can be maintained and multiplied in frozen embryo banks, and twins can be produced at will.

Enclosure. An area of land bounded by fences, hedges or walls.

Endemic. A descriptive term for plants or animals naturally occurring in a particular area. Also called indigenous.

Endoparasite. A parasite living on the inside of its host, e.g., Tapeworm.

Endosperm. Food storage tissue surrounding and nourishing the embryo in plant seeds. By the time the seed is fully developed the embryo may have completely absorbed the endosperm in some seeds, e.g., peas and beans, whilst in others part may remain and be absorbed after germination, e.g., wheat. Cereals are grown for the endosperm content of the grain, which is a valuable food source.

Endrin. A very toxic insecticide. A poly-chlorinated hydrocarbon.

Enteritis. Inflammation of the intestines.

Enterotoxins. Products of bacterial growth that cause inflammation of the intestinal tract, a form of food poisoning. Certain kinds of staphylococci can produce food poisoning owing to enterotoxins if they gain access to improperly refrigerated milk or soft foods such as custards, gravy and creamed meats. Enterotoxins are heat-resistant to the extent that they are not destroyed by pasteurization.

Enterozoon. Animal parasite which is inhabiting intestines.

Entire. An animal which has not been castrated. A stallion.

Entomologist. Pers on who specializes in the study of insects.

Entomology. Branch of agricultural science which is dealing with all aspects of insect activity including their structure, behaviour, their place in nature and form, whether they are harmful or useful *i.e.,* their economic importance. It also helps us to devise suitable control measures.

Entomophilous Plants. Plants which are pollinated by insects.

Environment (Ecology). External surroundings in which an organism lives. It is influenced and determined by the interactions of climate, water, light etc. and other living organisms. It affects the growth, development, and behaviour of the organism.

Enzootic Abortion. A virus infection of sheep affecting the placenta, causing abortion about 10-14 days prior to lambing time. Common in South-East Scotland and North-East England.

Enzootic Disease. An animal disease prevalent in particular areas or districts or during certain seasons (e.g., Enzootic Abortion) as distinct from epizootic diseases which spread rapidly and attack animals in large numbers almost simultaneously over large areas, even over entire continents (e.g., Foot and Mouth Disease).

Enzymes. Proteins present in living organisms which act catalytically to promote chemical changes whilst remaining unchanged themselves, e.g. Alpha Amylase.

Eosin. Tetrabromo fluorescein. A bacteriological stain.

Epidemic. The outbreak of an epizootic disease affecting great numbers of animals in one area at the same time and transmitted from place to place. The disease may not necessarily present in an area.

Epidemic Tremor. An infectious virual disease which mainly affects young chicken. The disease has been characterized by ataxia, muscular tremors and paralysis.

Epidemiology. Science dealing with development and or decline of an epidemic.

Epiphytotic. Means the sudden and destructive development of a plant disease on extensive area. It is known as epidemic in human disease and both the words have been used interchangeably by plant pathologists.

Epizootic. Enzootic Disease.

Epsom Salts. ($MgSO_4$. $7H_2O$). Magnesium sulphate, a white crystalline soluble salt, used often as a saline purgative, and as a Fertilizer, mainly as a foliar spray on fruit. It contains about 10% magnesium. (Kieserite).

Equilibrium Moisture Content. Moisture content of the product which is in equilibrium with the moisture content of the surroundings or hygroscopic equilibrium.

Equine. Pertaining to a horse or horses.

Equine Influenza. An equine disease which is caused by filterable virus. Respiratory syndrome is characterized by febrile reaction, nasal catarrh, depression and anorexia, while the abortion syndrome is characterized by high rate of abortion in pregnant mares.

Equine Malarial Fever. Infectious viral disease of equines which is characterized by a variable course and is usually associated with intermittent fever, progressive weakness, oedema and anaemia of a progressive or transitory type. It is caused by a filterable virus.

Equine Plauge. African Horse Sickness.

Equine Viral Rhinopneumonitis. Equine Influenza.

Equines. Horse and its relatives, the mule and the ass.

Equivalent Acidity. Pounds or kgs of pure calcium carbonate which is required to neutralize the acids produced in the soil from the quantity of fertilizer used, on the basis of which acid forming nature of fertilizer is decided; 1.6 lb in urea, 2.6 lb in calcium sulphate nitrate, 5.1 lb in ammonium sulphate.

Equivalent Basicity. In basic residue of any fertilizers measured in terms of calcium carbonate required to neutralize it. Term is used for expressing the basicity of a fertilizer.

Eradication. 1. The term used for the method of disease control in which the pathogen is eliminated after it has established 2. Absoluble destruction, by burning or deep ploughing etc.

Eradication Area. An area in which a programme designed to eradicate a particular animal disease is operating. Eradication schemes operating in such areas often involve the compulsory slaughter of animals or herds found, on testing, to have the specific. disease.

Ergosterol. Plant sterol which is acting as a precursor for vitamin D in animals, and is converted to this vitamin by ultraviolet radiation.

Ergot. 1. A fungal disease (*Claviceps purpurea*) of grasses, particularly cereals and especially rye. It attacks the seed heads and destroys the grain. Animals can suffer serious poisoning (ergotism) by eating diseased grain or products derived from such grain (e.g., bread, maize meal). In man it causes the condition known as St Anthony's Fire. 2. A small patch of horn, hidden by a tuft of hair, and located on the back of a horse's fetlock.

Erosion (normal). The wearing away of the land surface, particularly soil, by running water, ice and wind, etc. Ploughing up and down slopes can lead to gully formation, and light and friable soils can blow away in strong winds when exposed, usually in the spring. Wind erosion is a serious problem in the Fens.

Erosion (Animals). Gradual destruction of body tissues due to pressure, inflammation, irritation etc.

Erosion-Resistant Crop. A crop because of its dense foliage, root system etc. providing effective protection against soil loss by erosion.

Erosion Spalsh. Erosion which is preceded by the destruction of the crumb structure by rain-drops or drip.

Erysipelas. An infectious disease, mainly of pigs, in which hot, reddish inflammations of the skin and high fever occurs, and when severe lameness and difficult breathing follows. The disease commonly causes infertility and abortion.

Erythema. Reddening of the skin of swine which occurs due to congestion of the blood capillaries near the surface. Application of cold water or alcohol gives relief.

Escherichia coli. A bacterium found in the alimentary canal of most mammals. Certain strains cause disease, e.g., Coli Septicaemia in poultry and diarrhoea in various animals. Usually abbreviated to *E.coli*.

Espalier. A lattice-work, usually of wood, to train trees on. Also fruit trees so-trained, usually so that the branches grow horizontally.

Essential Amino Acids. Refer to those amino acids which are not synthesized in the body in sufficient quantities to support optimum metabolism.

Essential Amino Acids Index (EAAI). Method which is used for determining protein quality to which all the ten essential amino acids are considered. It may be defined as the geometric mean of the egg rotios of these acids to the food amino acids and may be calculated as follows :

$$\text{EAAI} = n\,\frac{100\,a}{a} \times \frac{100\,b}{b} \times \ldots \frac{100\,j}{j}$$

where *a, b, c...j* represent the percentage of essential amino acids in the food protein *a, b...j* percentage of same essential amino acids in the standard egg protein, and *n* represents the number of the amino acids.

This method is having the advantage that all the essential amino acids are considered but proteins having different amino acids composition may have the same index.

Essential Fatty Acids. Three unsaturated fatty acids linolenic acid ($C_{17}H_{29}COOH$) linoleic acid ($C_{17}H_{31}COOH$) and arachidonic acid ($C_{19}H_{31}COOH$) They are essential for the growth of mammals. As they are constituents of phospholipids and glycerides, they are vital for membrane production and fat metabolism. Only linoleic acid is needed in the diet as the other two fatty acids can be synthesized from it.

Estate. The total property in single ownership, often applied to the property of large landowners, including farms cultivated land, woodland, houses, etc.

Estivation. Dorment condition adopted by certain animals during summer. Compared with Hibernation.

Estrogen. Group of female sex hormones producing estrus.

Estrone. Female sex hormone occurring in the ovary, in the placenta and in urine of pregnant animals, women and plants etc.

Ether Extract. One of the fractions determined by the proximate analysis of animal feedingstuffs, containing mainly fats and oils together with various minor compounds, e.g., fat-soluble vitamins.

Ethylene C_2H_4. Growth hormone which regulates fruit ripening various aspects of vegetative growth, and is also important in the abscission process.

Etiolation. A condition of plants when grown in darkness, characterized by a general pale yellowness due to lack of chlorophyll, and by long internodes and small leaves.

Eutrophic. A descriptive term for nutrient or soil solution with nutrient concentrations optimal or nearly optimal for animal or plant growth.

Evaporated Milk. Milk thickened and enriched by evaporating much of its water content, but not sweetened like Condensed Milk.

Evaporimeter. An instruement which is used for the measurement of the amount of moisture which air will take up by evaporation. The quantity of water lost is usually recorded from an exposed pan (Open pan evaporimeter) or tank. Due allowance is made for water added by precipitation.

Evapotranspiration. The loss of water from an area (e.g. a field) by evaporation at the soil surface and by plant transpiration. Potential Evapotranspiration is the maximum transpiration possible under given weather conditions with a low-growing crop which shades the soil completely and has an adequate water supply.

Evergreen. A plant which keeps its leaves throughout the year. Mainly confierous trees and shrubs.

Evisceration. Of poultry and fish, refers to removal of the intestinal tract and other organs from the carcasses.

Eviscerated. Having had the viscera or bowels removed.

Ewe. An adult female sheep.

Ewe Hogg. Sheep Name.

Ewe Lamb. A female lamp under six months old.

Ewe-necked. A term for horses with thin hollow necks.

Excised-Embryo Test. This is used to test the germination of seeds of woodyshrubs and tree whose embryos need long periods of after-ripening before germination will occur. In this test, the embryo is excised from the rest of the seed and germinated alone. A viable embryo will either germinate or show some indication of germination, while a nonviable embryo becomes discoloured and deteriorates. The excision is carried out carefully to avoid any possible injury to the embryo. Procedures for germinating excised embryos are similar to those for germinating intact seeds.

Exochorion. The thick outer layer of the shell or chorion of an insect' egg. It is variously shaped and sculptured and having a small hole or micropyle at one end through which fertilization takes place.

Exoenzyme. An enzyme which is produced by a plant and exuded into the surrounding area where it breaks down potential food material which can then be assimilated by the plant.

Exostosis. Group of bone diseases of horses.

Exotic. A descriptive term for plants or animals introduced into an area or country from outside or abroad, not naturally occurring there.

Exotoxin. Substance having a pronounced toxicity and is produced by some pathogenic bacteria, which as soon as it is formed in the cytoplasm of the living pathogenic germ, filters out through the membrane and spreads in to the surrounding medium.

Expeller Process. A process for removing the oil from oilseeds in which the seeds are forced along a tapering tube, squeezing out the oil. The residues, containing usually less than 5% oil, may be used as animal feeds.

Extensive Farming. A method of farming in which a large amount of land is used to raise stock and produce crops, yields usually being about average as distinct from Intensive Farming.

Extracted Honey. Honey obtained by centrifuging uncapped broodless honey-combs.

External Seed-Borne Disease. Plant disease, in which the disease organism in its living state or its spores have been present superficially on the seed without any organic union with it, or as mycelium in superficial infection. It is controlled by application of fungicides to the surface of the seed.

External Seed Treatment. Treatment of seed for external seedborne fungal disease, by which fungicide like formalin sulphur, copper carbonate, organomercurials like Agrosan and ceresan, is throughly mixed with seed, preferably in a rotating drum so as to cover the seed surface with a layer of the fungicide. In the case of formalin, it gets diluted with an equal quantity of water and sprayed over a thin layer of seed and the seed heaped and kept covered with canvas overnight.

Eye. 1. An obsolete term for a brood, particularly of pheasants. 2. The seed-bud of a potato or an axillary bud on a plant.

Eyespot. A fungal disease (*Cercosporella herpotrichoides*) which attacks all cereals, causing oval or eye-shaped areas of black dots on leaf sheaths as they emerge from the soil in spring, and causing straws to weaken and lodge.

F

F. Class of flowers which are pollinated by butterflies and moths.

FAD. Flavin Adenine Dinucleotide. A coenzyme that is responsible for transfer of hydrogen form one compound to another.

FAO. Food and Agriculture Organisation which was set up by allied nations to serve as a clearing house through which the member nations can compare their agricultural consumption, production and trade programmes and suggest adjustment for meeting the world food needs.

F Class. Microplots having 4 rows 3.5 m long with rows 30 cm apart, and the plots 30 cm apart. They are used for testing the comparative merits of varieties under field growing conditions.

F_1 and F_2 Generations. Genetical terms for the offspring generations produced by a parental generation of plants or animals. F_1 is the first filial generation, and F_2 is the second filial generation (i.e., the offspring of F_1).

Factice. Rubber-like substance which is prepared by the reaction of such oils as linseed, soyabean and cottonseed with sulphur or certain of its compounds; a vulcanized oil.

Factor. A circumstance affecting the result of an observation or experiment. It is used especially for a set of related treatments in an experiment.

Facultative. An organism which has power to live under a number of certain specific conditions or to adopt an alternative mode of nutrition or life style of environmental condition etc., e.g., a facultative parasite may be either parasitic or saprophytic.

Facultatlve Anaerobc. Bacterium growing under either aerobic or anaerobic conditions.

Facutlative Clestogamy. Plant that can possess either chasmogamic flowers or those that do not open during pollination.

Facultative Parasite. Saprophytic organism having ability to become parasite.

Facultative Saprobe. Parasitic organism which is able to grow on dead organic matter.

Faeces. Solid waste or undigested material expelled by animals. Also called dung or manure.

Fairy Ring. An are or ring of lush, usually darker coloured grass in a pasture caused by fungi which grow in expanding circles in the form of a perennial, Mycelium. Such rings may be many years old and many yards across. The ring of lush grass is due to nitrate production, which acts as a fertilizer.

Fall. 1. An American term for autumn. 2. The dropping of an animal from its mother's uterus to the ground during birth. Also the number of offspring born in a herd or flock, etc., in a season.

Fallen Wool. A type of wool oddment purchased by the British Wool Marketing Board comprising wool which is either cliped from a dead sheep, or is badly shorn and is in the form of torn or broken pieces as distinct from an intact fleece.

Fallow. Land left unsown, usually for a season, during which it is ploughed and cultivated to kill perennial weeds by desiccation. The practice of fallowing is now less common.

Fallow Crop. A crop grown in well spaced out rows, e.g., potatoes or turnips, so that the land between the rows can be hoed and cultivated to control weeds. Also called a cleaning crop.

False Gid, False Staggers. A 'dazed' condition of sheep due to inflammation of the head affecting the brain, caused by maggots of the Sheep Nostril Fly feeding on the mucous membranes of the sinuses.

Family Farm. A farm run by a farmer and his family, usually with no other employees, and generally small to moderate in size.

Family Labour. Unpaid farm labour which is supplied by members of the family other than farmer himself.

Famine Commissions. Commissions which were set up by Government in 1880, 1898, 1901,. to ascertain the steps to be taken to ameliorate the severity of famines, with agricultural improvement as a subsidiary problem.

Fan. A basket used to winrow corn.

Fanged. A term for not caps having forked roots.

Fanning Mill. Machine which is operated by hand or electricity and equipped with a hopper, vibrating sieves, a fan etc. It is used for removing chaff, straw, weed seeds and dirt from grain.

Faraday's First Law of Electrolysis. Amount of chemical decomposition is proportional to the quantity of electricity which passes through the electrolyte.

Fardel-bound. Constipated. Especially used for cattle and sheep having food retained in the Omasum.

Farina. Ground corn or cereal meal.

Farinograph. Instrument which is used to measured the protein strength of flour so as to judge its baking quality.

Farm. An area of land, usually with a house and necessary buildings, used for agricultural purposes. The term originally meant such land leased or rented by a landlord and worked by another person, the farmer.

Farm Budgeting. Refers to the process of estimating costs, returns and net profit on a farm. A farm budget could be expressed by the following equation :

$$R = (Q_{yl} P_{yl} + \ldots + Q_{xl} P_{yn}) - (Q_{yl} P_{y} + \ldots + Q_{yn} P_{\alpha} n) - F \text{ or}$$
$$R \; \Sigma Q_y P_y - \Sigma Q_l P_\alpha - FC.$$

Where, Q_α represents quantity of a product consumed or removed from the inventory.

Q_y, quantity of a product sold or added to the inventory.

P_α price of the product consumed or removed from the inventory.

P_y, price of the product sold or added to the inventory.

R, total revenue (negative values represent the loss).

FC, fixed cost

and 1, 2, 3....*n* represent different items.

Farm Duty of Water. Seasonal quantity of water which is delivered to individual farm units under an irrigation project. Such duty is usually expressed in terms of acrefeet per acre, total depth of water over a unit area, or number of acres served per unit of flow. Such duty is able to express the actual rate of use of water at the farm, after all canal and other losses are eliminated, but includes losses in the farm ditches and also waste.

Farm Forestry. Refers to the practice of raising small woods on a farm in addition to normal cultivation for making the farm more self-sufficient in fuel, small timber, grazing facilities fodder and leaf manure, in addition to the other indirect benefits such as protection of crops against high winds and control over erosion. 10 to 15 per cent of farm area considered best are put under farm forests.

Farmer's Lung. A distressing respiratory condition caused by the inhalation of dust particles, *e.g.,* from mouldy feed or litter, containing fungal spores, which results in an allergic reaction in the air sacs of the lungs and causes breathing difficulties. Cows and horses can also suffer.

Farm Management. Branch of agricultural economics dealing with the business principles and practices of farming with an object of getting the maximum possible return from the farm as a unit under a sound farming programme.

Farm Map. Map of a farm. It depicts clear picture of the situation of blocks, plot or sub-plots, farm buildings, wells, undergound pipe lines, drainage, roads and bunds, fencing etc. It facilitates in planning the cropping scheme.

Farm Processing. Method or treatment which is used to prepare farm products for use or preservation. It is including the operation that maintains or improves the quality of farm products or changes their forms.

Farmstead. A farm house and the farm buildings associated with it.

Farmyard Manure. The faeces and urine of farm animals mixed with litter, mainly straw, to absorb the urine. This mixture is also called 'Long Dung'. Its composition is variable dependent on the animals contributing the dung, their diet and the kind of litter used. Farmyard manure is usually stored in manure heaps where bacterial activity releases ammonia, fermentation occurs, and the material generally degrades into simpler compounds. It becomesstructureless and the liquids drain as Dung Liquor. In this condition it is called 'Short Dung.' This material is spread on farmland to improve soil structure (being rich in Organic Matter) and also as Fertilizer, its main constituents being compounds of Nitrogen, Potassium and Phosphorus. Also called muck.

Farrand Test. A method for determining the Alpha Amylase content of Milling Wheat. Used as an alternative to the Hagberg Test. An extract of ground wheat is reacted with a solution of 'limit' dextrin (a derivative of wheat starch). After a measured time it is mixed with iodine solution, and the colour produced compared with that of a suitably diluted iodine 'limit' dextrin mixture. Loss of colour, due to Alpha Amylase in the wheat extract reacting with the 'limit' dextrin, indicates the activity of the enzyme, which is expressed as units. Sound wheat has an activity of 10 units or less. Wheat with an activity over 40 units is not normally accepted for milling. 40 Farrand Units is approximately equivalent to a Hagberg Falling Number of 160.

Farrier. A person who makes and fits horseshoes.

Farrow. 1. To give birth to a Litter of pigs. Also the litter itself. 2. A descriptive term for a cow or heifer not in-calf, or baren.

Farrowing Crate, Farrowing Rail. A steel crate used in a pig pen to contain a sow and prevent it from lying on and killing the piglets, whilst allowing them access to her teats. They are normally in the form of a long narrow rectangle, but are sometimes circular (the Ruakura pen). A farrowing rail serves the same purpose and is usually fitted a short distance form the pen walls and floor.

Farrowing Index. The number of Litters produced per year by a breeding sow.

Fat. 1. A term applied to animals reared for their meat (e.g., fat cattle, fat sheep) which have been fattened up and are in a ready condition for sale to a butcher at a market. 2. A dry measure of nine Bushels.

Fat Class. A visual appraisal of the degree of external fat development on a beef of sheep carcase, being one of the characteristics assessed in Carcase Classification. There are five main classes ranging from 1 (very lean) to 5 (very fat). For sheep, class 3 may be subdivided into 3L and 3H; whilst for beef, classes 4 to 5 may be subdivided into 4L and 4H, and 5L and 5H.

Class	Percentage Fat	
	Beef Carcases	Sheep Carcases
1	< 4.5	< 6.0
2	4.5-7.4	6.0-9.9
3L	7.5-8.9	10.0-11.9
3H	9.0-10.4	12.0-13.9
4	10.5-13.4	14.0-17.9
5	> 13.4	> 18.0

In the case of pigs, fatness is measured with an Intrascope which indicates fat depth to the nearest millimetre.

Fat Content. Amount of butterfat in milk, usually stated in percentage.

Fat Hen. A weed (*Chenopodium album*), particularly of arable land,

Fatling. A young animal fattened for slaughter.

Fats. Storage materials found in living organisms, mainly in liquid form (oils) in plants, and in solid form (also called adipose tissue) in animals, comprising mixtures of glycerides (condensation products of glycerol and Fatty Acids). Most animals deposit fat in the body as an energy store. Some plants, mostly tropical, produce seeds rich in oil which is extracted mainly for cooking, the residue being used for cattle Cake.

Fatstock. Livestock fattened for sale in a market.

Fat Soluble Vitamins. These are usually found associated with the lipids of natural foods e.g., vitamin A, D, E and K.

Fatty Acids. Organic acids, thirteen of which occur in natural Fats. They are either (a) saturated (i.e. molecules to which no further atoms can be added), found mainly in solid form in Animals Fats, or (b) unsaturated, found mainly in liquid form (oils) in plants. The most common fatty acids in natural fats are palmitic, oleic and stearic. (Steapsin)

Fauna. The population of animal life in a place or area, or during a particular period.

Faverolle. A heavy table breed of poultry, with a single comb and slightly feathered white legs bearing five toes, producing white eggs. Variously coloured including blue, buff, salmon and white.

f.c.s. An abbreviation for 'filled, carted and spread.' An agricultural valuers' expression for the labour required to transport manure from the farmyard to the fields. Now rarely used, the expression 'labour to farmyard manure' being more common.

Feather. 1. One of the dermal outgrowths forming the plumage of a bird. 2. Long hairs found on the sides and on the back of the legs of some breeds of horse e.g. the Shire.

Feathered Maiden. A Maiden tree which has grown side shoots.

Fee. An obsolete term for livestock, particularly cattle.

Feed Block. A block of foodstuffs left on pastures, particulary in hill areas, for sheep to lick at will as an aid to maintaining Condition. Normally they contain a Carbohydrate source (e.g., cereals, glucose, or molasses), Protein and Urea (sometimes excluded), Minerals, Vitamins and Trace elements.

Feed Off. The practice of allowing a crop to be eaten by animals whilst still in the ground.

Feed Ring. A railed circular container mainly used for hay and straw, from which livestock feed around its circumference.

Feedingstuffs. The various foods available for farm animals, loosely classified into four main groups, viz, (*a*) Green Forages, (*b*) succulent foods, mainly roots and tubers, e.g., turnips, potatoes etc., (*c*) dry roughage or coarse fodder, e.g., hay derived from grasses, cereals and legumes, or Straw derived from cereals and the haulms of peas and beans, and (*d*) Concentrates. Also called feeds.

Feedlot. An area of land in which animals (very commonly beef cattle) are accommodated at very high density. The feedlot does not contribute at all to the production of animal feed, all of which has, therefore, to be brought to the animals from outside the feedlot.

Feer. To plough the first furrow when starting ploughing.

Fehling's Reagent. Refers to a solution of cupric sulfate, sodium potassium tartrate, and sodium hydroxide. It is used to test for the presence of reducing compounds such as sugars.

Fell. 1. The hide of skin of an animal, particularly when covered with hair. A fleece. 2. A term mainly used in Northern England for mountainous moorland and hill pastures. 3. To cut down a tree.

Fellmonger. A person who prepares animal skins for tanning.

Felloe' Felly. The circular rim of a wheel to which the spakes are attached.

Fen. A lowlying, marshy area of land, with a Peat soil which may be alkaline to slightly acid, as distinct from the very acid soil of a Bog.

Fenland Rotation. A three-course system of crop Rotation developed on the deep, rich, silts and Fens of East Anglia, in the sequence potatoes-sugarbeet-wheat. The short sequence increases the likelihood of attack by pests and disease, and other high value crops are frequently included, such as rootcrops, brassicas, bulbs, peas and celery.

Ferm. A farm.

Fermentation. The breakdown of organic substances by the action of Enzymes, usually secreted by living organism such as Bacteria and Yeast. Heat and gas (e.g., carbon dioxide) are usually evolved. The principle of fermentation is used in the making of Silage, wine, vinegar, etc.

Fermented Milk. Also known as cultured buttermilk. Milk seeded with cultures of lactic acid bacteria to bring about souring and to cause changes in texture, aroma and taste. A refreshing beverage.

Ferret. An albino variety of the polecat, half-tamed and used to catch or unearth rabbits and rats.

Ferruginous Soil. Soil having large amounts of iron minerals, especially limonite and hamatite. It is characterized by the colour or shades of yellow and brown.

Fertile. 1. A descriptive term for plants or animals able to reproduce successfully, as distinct from being sterile. Also those able to produce offspring prolifically. Used also to describe eggs capable of producing a chick, or seeds capable of Germination. 2. A descriptive term for land or soil, rich in nutrients, and capable of producing good crops.

Fertilization. 1. The enrichment of soil by the application of Fertilizer. 2. The union of male and female gametes in reproduction to produce a zygote.

Fertilizer. In general terms anything added to the soil to increase the amount of plant nutrients available for crop growth. Normally applied to inorganic chemicals and non-bulky organic substances (i.e., excluding Compost, Farmyard Manure, etc.,, which are commonly called Manures). Fertilizers basically provide the major plant nutrients (*i.e.,* Nitrogen Phosphorus and Potash), either individually (Straight Fertilizer) or as a combination of two or three (Compound Fertilizer)., sometimes together with Trace Elements.

(*a*) The nitrogenous fertilizers include Ammonium Nitrate, Nitrate of Soda, Sulphate of Ammonia, Anhydrous Ammonia, Urea, Ammonium Phosphate and Liquid Nitrogen Fertilizers. (*b*) The phosphatic fertilizers may be quick acting (water soluble)) or slow acting (water insoluble) and include Ammonium Phosphate, Superphosphates, Bone Flour, Bone Meal, Fish Guano and Reddzlaag. (*c*) The Potash fertilizers include Muriate of Potash, Sulphate of Potash, Kainit and Nitrate of Potash. Fertilizers applied to the soil usually quickly dissolve in the Soil Water and are taken up by plant roots or absorbed by soil colloids (Cation Exchange). Excess fertilizer may be leached through the soil into the Ground-Water. Fertilizers now-a-days are mostly produced in easily stored granulated form but powders and crystals, and also less concentrated liquids are available. Inorganic chemicals such as Limestone and Gypsum applied to correct such soil conditions as acidity and salinity may also be considered as fertilizers.

Fertilizers Distributors. Machines which distribute Fertilizer on the land usually by broadcasting if in a solid form (i.e., crystals, granules or powders), or by spraying if in liquid form, and sometimes by injection into the soil (e.g., Anhydrous Ammonia). Broadcast types comprise a hopper which feeds the fertilizer to a spreading mechanism, the most popular being either a spinning disc or oscillating spout, which distribute fertilizer in swaths of variable width. These machanisms are less accurate than the 'full width' distributors such as the once popular 'plate and flicker' type or the more modern agitator or pneumatic type spreaders. Fertilizers can also be applied by a Combine Drill, Placement Drill, and by aircraft as top dressing, particularly to cereals.

Fertiliser Grade. An expression indicating the percentage of plant nutrients in a fertilizer *e.g.,* a 12-5-5 grade of fertilizer means 12 per cent nitrogen, 5 per cent phosphoric acid (P_2O_5) and 5 per cent potash (K_2O).

Fertilizer Ratio. Referring to the relative proportion of three major plant nutrients, keeping the percentage of nitrogen as one in the ratio. Thus a 6-12-6 fertilizer mixture having 6 per cent nitrogen, 12 per cent available P_2O_5 and 6 per cent soluble potash has a nurtrient ratio 1:2:1.

Fertilizer Requirement. The quantity of particular plant nutrients required, in addition to those already contained in the soil, to increase crop growth to a desired optimum.

Fertility Gradient. Means the variation in nature fertility in any direction across an area of ground.

Fertility (Soil). Means the ability of a soil to supply all the essential nutrients in optimum amount in a form that is readily available to plants.

Fetlock. The tuft of hair which grows above a horse's hoof or the part of the leg where it grows.

Fibre. A constituent of animal Feedingstuffs consisting mainly of Cellulose and Lignin, which aids digestion by stimulating muscular activity in the digestive tract. It also facilitates the digestion of concentrates by opening them up to the action of the digestive juices. All farm animals require a certain amount of fibre except when very young and on a liquid diet. Ruminants require the most fibre and derive energy from it by bacterial digestion in the rumen. Cows require adequate fibre to produce milk with a high percentage of Butterfat. Sheep and horses require less fibre than cattle but more than pigs. Excess fibre in the diet can prevent an animal eating sufficient food to derive adequate essential nutrients. Hay and and Straw are characterized by a high fibre content.

Ficundity of Insects. The average numbers of eggs laid by an insect has been from 100-200 but in some particular cases, especially the social insects, there may be more than this, *e.g.*, a queen may lay 2,000 eggs a day; a queen termite may lay 60 eggs a minute with total of several millions.

Fiddle. A small portable device carried on the shoulders for broadcast sowing, especially grass and clover seeds in hilly areas. It comprises a hopper which slowly drops seed on to a ribbed disc which the operator causes to rotate and throw the seed by pushing and pulling an attached 'bow'. Very rarely used.

Field. An area of land, usually enclosed by a ditch, fence, hedge, etc., for cultivation or grazing purposes. Fields are usually sown to a single crop at one time. In recent years the trend has been the removal of field divisions (*e.g.*, hedges) to facilitate cultivation and increased efficiency, and the larger fields created are sometimes termed management units. Such larger fields also allow cropping flexibility and sometimes several different crops are grown in the field at the same time.

Field Capacity. The state of soil when all the soil moisture that is able to drain away freely has done so. The remaining moisture is held by the forces of surface tension around soil particles and in capillary pores (Capillary Water).

Field Capacity Zone. Buffer zone on which water has been held against gravity under free drainage. It is this zone which does not allow the moisture and consequently the salts from moving in a vertically upward direction.

Field Crops. Herbaceous plants which are grown in cultivated field under more or less extensive system of culter.

Field Duty of Water. Quantity of water which is given to individual fields or orchards under an irrigation project. It has been expressed in terms similar to those for farm duty of water.

Field Efficiency (Plough). Refers to the ratio of effective field capacity, and theoretical field capacity. It is expressed in percentage.

Field Irrigation Requirement of a Crop. Means water which is required by crops exclusive of rainfall and contribution from soil profile.

Field Moisture. Refers to the per cent moisture content in a field sample of soil at any one time.

Field Moisture Deficiency. Quantity of water needed to restore the soil moisture content to field moisture capacity.

Field Moisture Equivalent. Minimum moisture content which is expressed as a percentage of the weight of the oven-dried soil at which a drop of water kept on a smoothed surface of the soil will not immediately be absorbed by the soil but will spread out over the surface and give it a shiny appearance.

Field Waste. That portion of the irrigation water which has not been absorbed or evaporated as it passed over the irrigated land.

Field Room. The time needed for a crop stooked (Stock) in a field to dry sufficiently for stacking.

Field Sports. Sports carried out in the countryside such as fóxhunting, hare coursing, horse racing, etc.

Fieldman. A farmworker, the major part of whose work is in the fields, as distinct from around and in the farm buildings.

Filbert. The nut of the cultivated Hazel.

Fill. 1. The contents of the digestive tract of an animal. 2. The space between the shafts of a cart in which a horse stands when harnessed, the horse itself being the filler.

Filler. A tree temporarily growing in between fruit trees in an orchard until they mature and require the space.

Filly. A young female horse between 1 and 4 years of age.

Filly Foal. A female horse up to 1 year of age.

Film Yeast. Type of yeast (microscopic fungs) growing on the surface of certain nutrient liquids and producing heavy folded films. Such yeasts are widely distributed in nature. They are generally found in fermentation process on the surface of sugar containing solutions, pickle brines, or other acid and sugar containing liquids. These yeasts generally grow in very high concentrations of salt, acid, and sugar.

Filter (Bacterial). The term used for a special type of filter through which bacterial cells fail to pass. A filter is used for separating microorganism from a gas or liquid in which they become dispersed. Filtration is a convenient method to sterilise heat labile liquids and solutions which cannot be autoclaved. This is carried out merely by passing the virus free liquid through a suitable filter.

Filterable Virus. Microorganism which is capable of passing the pores of microbiological filters (or a bacterial filter).

Fingers. Units with flat faces (known as ledgers projecting from and bolted to the cutter bar of a Binder, Mower or Combine Harvester, across which the knives slide backwards and forwards and cut the grass blades or crop against the fingers.

Finish. An Open Furrow being the last one ploughed in a field.

Finished. A term applied to animals, particulary beef cattle which have been carefully fattened and are ready for sale at market, having an acceptable degree of fat cover. Finished animals are expected to produce good quality carcases.

Fire Blight. A notifiable Disease caused by bacteria (*Erwinia amylovora*), mainly of pears and apples, giving a scorched appearance to the fruit. Spread by insects.

Fire Curing. The term used for the process of curing tobacco specially for jaffina tobacco used for chewing purposes. In this process the leaves are wilted for a few hours in the field tied into bundles and hung on laths in a smoke hut then smoked for 12 hours by burning.

Firebreak. An existing barrier which is natural or, otherwise, or one prepared before a fire takes place from which all or most of the inflammable materials have been removed. It is designed to stop light surface fires and to serve as a line from which to work and counter-fire if necessary.

Fire-calf Heifer. A term used in some areas for a Heifer after having borne its first calf but before it has borne a second calf.

First Cross. The progeny from the mating of a male and female from separate pure breeds of animals or varieties of plants.

First Service Conception Rate. The percentage of cows in a herd in which pregnancy is established after the first service.

Fish Ladder. An inclinded flume which is carrying water from above to below a dam, at a velocity against which fish can easily swim.

Fish Farming. The breeding and rearing of fish in tanks, ponds, cages, etc., ultimately for harvesting and slaughter for the table.

Fish Gauno. A phosphatic fertilizer made from unmarketable fish and fish waste products after the extraction of variable composition. The nitrogen content is usually 7-8% and the phosphoric acid content 4-8%. Usually only used for horticultural purposes. Also called fish meal.

Fish Manure. The unprocessed offal from gutted fish and waste or unsold rotting fish, spread on fields as Manure.

Fish Meal. An animal Feedingstuff consisting of dried and ground waste or unsold fish and fish filleting residues, by law containing not more than 6% oil and 4% salt. White fish meal is derived from non-oily 'white fish (*e.g.*, cod, haddock, etc.) and usually contains about 66% protein. Other fish meals are derived from oily fish (*e.g.*, herring, pilchards, etc.) after most of the oil has been extracted and are usually richer in protein than white fish meal. Fish meal is mainly fed to pigs and polutry, but is also included in rations for dairy cows, calves and other farm animals. Fish Guano.

Fish Pond. (1) Pond having fish (2) Pond natural or artificial where fish life has been supported by the sewage discharged therein.

Fish Pond Fertilizer. Inorganic or organic matter which is applied to a fish pond for promoting the growth of plankton and aquatic plants on which the fish depend directly or indirectly for food.

Fish Screen. 1. Refers to a screen which is placed across the head of, or in, an intake canal or pipe line, to prevent fish from entering. 2. Barrier placed across the inlet or outlet of a pond for preventing the passage of fish.

Fishway. Structure with pools and drops which is usd to facilitate the migration of fish around dams or other obstructions in streams. They are needed on many streams where fish periodically migrate upstream to the head waters to spawn, the young fish coming downstream again. One form is known as fish elevator, another form fish ladder.

Fives. A term for samples taken from hop Pockets (for assessing type) whose identity numbers end in 5 (*i.e.*, 5th, 15th, 25th, etc.). Samples are thus taken from one pocket in ten.

Fixation of Nitrogen. Refers to the fixation of atmospheric nitrogen in the soil by symbiotic and non-symbiotic bacteria.

Fixation of Phosphorus. Conversion of soluble phosphorus nutrient in the soil into less soluble and unavailable forms.

Fixation (Photosynthesis). Process of producing carbohydrates by adding hydrogen to carbon and oxygen.

Fixative. Reagent which will bring about fixation.

Fixed Ammonium. Ammonium which is held in a soil and is neither water soluble nor easily exchangeable.

Fixed Capital. The capital invested in land buildings and relatively permanent fixtures.

Fixed Ground Water. Ground water which is held in saturated material with interstices so small that it gets permanently attached to the pore walls or moves to slowly that it is usually not available as a source of water for pumping.

Fixed Moisture. Moisture which is held in the soil below the hygroscopic limit.

Fixed. Breed In.

Fixed Equipment. In general terms anything on a farm, excluding growing crops, which cannot readily be moved.

Fixed Phosphorus. 1. *Phosphorus.* It is changed to less soluble form because of reaction with soil; moderately available phosphorus 2. *Applid Phosphorus.* Phosphorus which has not been taken up by plants during the first cropping year 3. *Soluble Phosphorus.* Phosphorus which gets attached to the solid phase of soil in forms highly unavailable to crops 4. *Unavailable Phosphorus.* Phosphorus other than readily or moderately available forms.

Fixed Potassium. Potassium which is held by a soil and is neither water-soluble nor readily exchangeable.

Flag Leaf. Last fully expanded leaf in a cereal plant.

Flail. 1. A hand implement used to thresh corn, comprising a wooden beating bar or swingle, hinged to or loosely attached (often by leather) to a handle. Also called a dreshel or a Stick and a Half. 2. One or more free swinging arms on certain machinery (*e.g.*, Flail Forage Harvester, Flail Hedger) which, when rotated at speed, sever vegetation.

Flail Forage Harvester. A type of Forage Harvester consisting mainly of one or more swinging arms or flails, each bearing a number of knives, and attached to a horizontal high-speed rotor. The flails are covered by a hood from which a chute delivers the cut forage to a trailer. Flail forage harvesters may be trailed or semi-mounted on a tractor, and offset versions of both types exist which avoid tractor wheel damage to the crop.

Flail Hedger. A type of Hedge Cutter commonly used now-a-days for regular trimming. They are usually driven hydraulically or mechanically by p.t.o. from a tractor and consist of a high-speed cylinder bearing free swinging flails.

Flaked Maize. A highly digestible, starch-rich feedingstuff derived from maize which has been steam treated, rolled and dried. It has a relatively low oil content and is often given to pigs.

Flank. The side of an animal between the ribs and the thigh.

Flash Pasteurisation. Term includes all short time heat treatment methods which are not scientifically controlled. It was formerly used for milk destined for manufacturing purposes, but has now been almost entirely replaced by the accurately controlled HTST process.

Flat Rate Feeding. A system of feeding by which cows are fed a standard fixed quantity of concentrate per day, from calving to turnout, thereafter depending on grass alone. A variation commonly used with autumn calving cows is stepped feeding, in which the level of concentrate fed to groups of cows in the herd is reduced in one or two steps through the winter, with concentrate feeding being discontinued on turning out to grass. The success of these systems depends on the cows having unrestricted access to good quality forage, usually grass silage. If necessary, the silage intake can be reduced by supplementation with other suitable feedingstuffs, *e.g.,* Brewers' Grains, Dried Grass.

Flat Roll. A type of Roller with 2 or 3 smooth surfaced cylinders mounted on an axle.

Flavoured Milk. Type of milk which is standardized to a certain fat percentage with skim milk to which are added flavouring materials, chocolate, fruit syrups, vitamin concentrates, etc.

Flax. Varieties of the plant *Linum usitatissimum,* selected and grown for their long stems rich in fibre, used in linen manufacture.

Flae Beetle. A small dark beetle (*Phyllotretra sp.*) which causes extensive damage to Brassica seedlings, eating small holes in the cotyledonous leaves (Cotyledon), especially during hot dry weather.

Fleam. A veterinary instrument for bleeding cattle.

Fleck. A fat layer surrounding the kidneys of pigs. Also called flare.

Fleece. A sheep's coat of wool. Also the wool shorn off a sheep and maintained on piece.

Flemish Gaint. A very large breed of rabbit, dark grey in colour with white specks.

Flick. Rabbit fur. Also waste rabbit fur, used as nitrogenous Fertilizer.

Files. A general term for winged insects, but particularly applied to the two-winged insects of the order Diptera, the mouth parts of which form a proboscis for piercing and sucking, especially blood. Many are important in agriculture, causing or transmitting disease to plants and animals. Some are parasitic (*e.g.,* Cleg, Gadfly, Ked, etc.) whilst the Larvae of others attack livestock and crops (*e.g.,* Blow Fly, Crane Fly, Frit Fly, Warble Fly, etc.). Clouds of flies, especially when biting, make cattle very restless.

Flinty. Used to describe cereals when very ripe and hardened like flint.

Flitch. A side of bacon, salted and cured.

Float. A low cart used to carry livestock or produce.

Flocculation. The aggregation of small particle to form larger ones. Particularly applied to soils and the formation of crumbs (Deflocculation).

Flock. 1. A collection or company of animals. Particularly applied to sheep, goats and birds. 2. A locks or tuft of wool (and cotton or hair), especially when waste and used to stuff upholstery, mattresses, etc.

Flock Book. A record for a particular breed of sheep, maintained by the Breed Society, of registered sheep conforming to the breed type and their pedigree relationships.

Flok Mating. Allowing several cocks to mate at will with the hens in a flock of poultry, as distinct from using only one cock run with the hens in a pen (pen mating).

Flockmaster. The owner or person in charge of a flock of sheep.

Flora. The plant population of an area or country. Also a botanical ordering of those plants together with descriptions for identification.

Flue Curing (Tobacco). Process which involves three distinct stages *viz. yellowing, fixing the colour* and *drying of the leaves.* During yellowing, temperature is kept at 90°-95°F for about 24-40 hours with high humidity.After yellowing, temperature is raised gradually and the humidity of the barn lowered by operating ventilators to get fixation of the yellow colour, after this the temperature is raised to 160°F to dry the midribs. This completes the process of curing.

Flue Dust. A fertilizer containing 5-15% Potash derived from industry, mainly as waste products from iron and steel manufacturing processes.

Flukes. Parasitic flatworms found in the liver, guts, lung and blood vessels of animals. Liver Flukes are especially common in sheep and cattle.

Flukey Area. A badly drained pasture where mud-snails are found, which are the alternative host in the transmission of liver fluke disease.

Fluorosis. Also known as fluorine poisoning. Chronic and insidious disease of cattle which occurs due to prolonged ingestion of fluorine. Characterized by stunted growth, lameness, mottled and irregular teeth, thickening of jaws and bones of limbs.

Flush. 1. A sudden, generally rapid growth of foliage, usually applied to grass. 2. To give an animal a richer diet prior to mating in order to improve its condition.

Flute Budding. Refers to a method of budding where in the bark is removed from the root stock almost completely encircling it, leaving a narrow connection (about 1/8 of the circumference) between the upper and lower part of the stock. It is replaced by a similar bark having an active bud from a desired variety. The narrow connecting strip of bark on the stock keeps the top of the stock alive, if the bud patch does not unite.

Fly Wool. Pieces of loose wool that fall from a fleece during shearing. (Wool Oddments).

Flying Bent. A coarse tussock-forming grass found in wet habitats on Moorland, Heath, Fens and Marshes, with purple flowers eaten only by sheep. Also called Purple Moor Grass or Purple Melick-Grass.

Flying Flock. A flock of sheep temporarily imported onto a farm, normally for less than a year, and then sold out.

Foal. 1. A young horse in its first year. A colt foal is an uncastrated male foal, and a filly foal is a female foal. 2. To give birth to a foal.

Fodder. Food supplied to livestock, particularly dry roughages such as Hay and Straw. Sometimes used loosely to mean Forage.

Fodder Beet. A root crop of the genus *Beta* which also includes sugar beet and mangel; usually grown following cereals and used to feed stock. It is intermediate between mangel, and sugar beet, having a higher Dry Matter and sugar content but smaller root than mangel, but a lower dry matter and sugar content and larger root than sugar beet. Fodder beet may be classified into medium dry matter (14-18%) and high dry matter (19-20%) varieties.

Fodder Crop. A crop grown for use as animal feed, *e.g.*, Hay Kale, Rape, Lucerne, Trifolium, etc., and usually consumed in the green state. Also called green crop or forage crop.

Fog. Pasture grasses allowed to grow during late summer and autumn, providing winter grazing for sheep and cattle. Land left with such grass on it. Also to graze such grass in winter.

Foggage. Synonymous with Fog. Also used to describe rough coarse grasses of mountain and hill areas and Rough Grazing.

Fold, Folding. An enclosure or pen for sheep, usually movable and made with hurdles or wire. Usually used to enclose sheep (and also pigs) in a field with a growing arable crop (*e.g.*, turnips, kale), moved each day to another part of the field or enlarged once the crop in that area has been consumed. This practice is termed folding.

Fold Units. Mobile units for poultry consisting of a house comprising a sleeping area and nest box attachment together with a run. Usually designed to accommodate about 20 laying hens, and moved regularly to provide clean land for the hens to spread the faeces evenly over the land.

Followers. 1. Young cows in a dairy herd not yet in milk and still maturing to replace their mothers. 2. Cattle put out to graze a pasture more thoroughly after a dairy herd has taken the best grass.

Food and Agricultural Organisation. An autonomous agency of the United Nations which aims to raise nutritional levels and living standards, particularly in poor countries, to improve the production and distribution of food and agricultural products, and to better the condition of rural populations, there by ensuring freedom from hunger in the world and contributing to the expansion of the world economy. F.A.O. employs over 3000 professional planners and technicians working on a wide variety of projects in many parts of the world.

Food Conversion Ratio. The number of kgs of food consumed by an animal required to produce a liveweight gain of 1 kg. A measure of the efficiency of the animal in converting food into flesh, a small ratio indicating high efficiency. Also called feed conversion ratio.

Food Poisoning. General term which is applicable to all stomach or intestinal diturbances due to food contamination with certain micro-organisms or their toxins.

Food Unit. 1 kg of average barley, equivalent to 0.71 kg starch equivalent. Also called a fodder unit.

Foot and Mouth Disease. A highly infectious viral notifiable disease of cattle, pigs, sheep and goats, characterised by the development of blisters in the mouth, causing considerable salivation, and on the feet resulting in lameness. Death is not usual, but animals cease gaining weight and milk production in dairy cows drops. Epidemics occur from time to time and the disease has been endemic in many parts of the world. The disease can be spread by infected bones, offals, birds, etc., and by the wind. Diseased animals and their contacts are generally slaughtered.

Foot Rot. A chronic wound-infection which has been characterized by extreme lameness and frequently by distortion of the hoof. It occurs due to irritation by mud which dries and cracks the skin, by sand and by gravel between the claws. Animals badly affected may walk on 3 legs.

Forage Blower. A machine used to fill a Silo, particularly of the tower type, with finely chopped material for making Silage, by blowing it up a duct, often to a height of 15 m (50ft) into the silo. Some kinds of forage blower also chop the crop, these being called cutter blowers.

Forage Box. A large mobile container, usually with wheels, similar to a Dump Box, used to transport Forage from a Silo to a Manager. The forage is discharged in a steady even flow, usually by a transverse moving floor mechanism or elevator belt, driven by p.t.o. from a tractor. Also sometimes used to transport forage from the field to managers for herds under Zero-Grazing.

Forage Crop. Crops which are grown primarily for livestock feed, to be either harvested for hay, silage or green feed or harvested by grazing animals.

Forage Harvester. A machine which cuts, chops and loads (by chute) green crops (*e.g.*, grass, lucerne, etc.) into an adjacent or towed trailer, for subsequent silage making. Sometimes the crop is precut and allowed to wilt in windrows before being collected by a pick-up reel substituted for the cutter mechanism. There are three main types, viz., Double Chop Harvester, Flail Forage Harvester and Metered Chop Harvesters. They are mainly trailed or semi-mounted but self-propelled machines are available.

Foremilk. The first few squirts of milk drawn from a cow's udder by hand into a strip cup before starting milking. This milk has a high bacterial content.

Forest. A large area of uncultivated land covered by trees and underwood. Blocks of woodland in excess of 1000 acres are classed as forest by the forestry commission. The original meaning was unenclosed woodland or open, mainly treeless, areas reserved for hunting, usually belonging to the Crown, *e.g*, the New Forest.

Foremalin. An aqueous solution of formaldehyde used as a sterilant, mainly of glasshouse and potting soils and implements, and occasionally used as a disinfectant.

Fors. Hairs in fleece, mixed with the wool. The removal of such hairs from wool is called forsing.

Forward. A descriptive term applied to both livestock and crops that are more advanced in their development at a particular time than normal.

Foul Brood. A bacterial disease (*Bacillus sp.*) of bees which kills the larvae and causes their decomposition.

Foul-in-the-foot. A bacterial infection (usually *Actinomyces necrophorus*) of the hooves, particularly of the hind feet, of cattle, normally around wounds or cuts. Swelling, heat, pus production and lameness occur. Muddy fields and gateways are particular sources of infection.

Four-tooth Sheep. Sheep of about 18-21 months of age.

Fowl Cholera. A contagious bacterial disease (*Pasteurella apisepticus*) of poltry characterized by sudden high fever and profuse green diarrhoea.

Fowl Paralysis. Marek's Disease.

Fowl Pest. A term mainly used for Newcastle Disease. Also another name for fowl plague, a similâr viral disease of poultry which is rare in the British Isles.

Fowl Pox. A viral disease of fowls causing warts on the comb, wattles and other parts of the head.

Fowl Typhoid. An infectious bacterial disease (*Salmonella gallina rum*) of poultry, mainly affecting Pullets, causing drowsiness and loss of appetite.

Fox. A carnivorous canine predator (*Vulpes vulpes*) which is known to take young lambs and poultry and is a vector of rabies.

Foxy Hops. Over-mature dried hops with a reddish-browncolour. The colour may also be caused by disease or decay.

Frame Harrow. A type of harrow with an articulated sectional frame and no wheels, providing flexibility to follow the contours of the land.

Free-grazing. The most common method of grazing stock with few, if any, controls. Stocking rate is usually adjusted to the supply of herbage but over-and under-grazing often occurs. Fencing costs are kept to a minimum and intensive stock supervision is avoided. (Grazing systems).

Free-martin. A heifer calf born as a twin with a bull-calf, usually sterile.

Free-range. A system of keeping poultry in which the birds are provided with small houses and are allowed to run free over a field or large enclosed area. Each house is usually designed to accommodate 50-150 birds and is raised slightly above the ground. Most poultry are now-a-days kept under more intensive systems such as battery cages.

Pigs may also be kept under a free range system, with breeding sown provided with cheap mobile housing or arks for protection, and having the run of an area of grass. The majority of pigs are also kept under more intensive systems.

Freshen. A term applied to a cow as it approaches calving.

Friesian. A breed of black and white cattle originating in Holland and exported across the world, so that some variations now exist. The British Friesian is intermediate between the well-muscled, short-legged, dual-purpose Dutch Friesian, and the Holstein Friesein (Holstein) of Canada which produces a high yield of milk and Butterfat. It is characterised by longer legs than the Dutch Friesian, with a deep body, long and wide hindquarters and a large udder. The average annual Milk Yield in 1981/82 (milk yields) was 5885 kg (5703 litres) with a 3.83% butterfat content. Red and white types are also known.

Frit Fly. A small shiny black fly (*Oscinella frit*) that lays its eggs on oat seedlings and tillers, which are then attacked by the larvae. The next generation of files lay eggs in the glumes and the larvae attack the developing grain. Wheat and maize may also suffer from frit fly attack.

Front Loader Arms. Hydraulically operated arms fitted on either side of the front end of a tractor to which various implements may be attached.

Fruit Fly. 1. Small yellow-brown flies (*Drosophila sp.*) which feed on fruit. 2. Large flies, *e.g.*, celery fly, gall flies, etc., whose larvae feed on fruits.

Fruit Sugar. Fructose.

Full Mouthed. A descriptive term for livestock in which a complete set of permanent teeth have grown, *e.g.*, after 3 years in sheep.

Fumigate. To destroy bacteria, insects and other pests by exposing the place they are infecting *e.g.*, a farmbuilding, to poisonous gas or smoke (Disinfection).

Fungicide. A chemical used to destroy fungi and thus control fungal diseases.

Fungus. One of a large group of plants including moulds, mushrooms and yeasts, characterized by lack of chlorophyll. They are either parasities, many of which cause crop disease (*e.g.*, Potato Blight, Mildew, Ergot, Rust) or Saprophytes which are important in the release of nutrients to the soil from dead plants and animals. They are mainly constructed of thread like hyphae and are classified into three groups, vis., Ascomycetes, Basidiomycetes and Phycomycetes.

Furrow. The groove or trench in the soil made by a plough as the mouldboard turns over the soil in a more or less continuous strip (the furrow slice). A completely ploughed field, before other cultivations are undertaken, is said to be 'in the furrow'.

Furrow Opener. Another name for a Coulter.

Furrow Press. A heavy type of multi-ring Roller usually trailed behind a plough, with wedge-shaped rims to some of the wheels (the press wheels) which compress the lower parts of the furrow slices. Sometimes used when preparing light land for wheat following a Ley. Also called press.

Furrow Wheel. A wheel on a trailed plough which runs in the previously turned furrow.

Fym. An abbreviation for Farmyard Manure.

G

Gadfly. A large, dark, blood-sucking fly (*Tabanus sp.*) with clear wings and large brilliant eyes. Gadflies attack livestock between May and September, normally on warm sunny days, and their biting can cause distress.

Gadsman. A Scottish term for a person who drives horse-drawn ploughs.

Gait. 1. A term used in some cases for a sheaf of corn set on its end. 2. The manner of walking, running, trotting, etc., of an animal.

Galactorrhea. The term used for an excessive secretion of milky, fluids from the breast, especially when not associated with pregnancy.

Galia or Sand Leaves. Lowermost 3 to 4 leaves of inferior quality in tobacco plant. These constitute about 10 to 15 percent of the total bulk.

Gall. 1. An abnormal growth of plant tissue, usually in response to parasitic infection, mainly by a variety of Insects (e.g., certain wasps, saw-flies, midges, Aphids, and some moth and beetle Larvae) and Mites, the offspring of which are provided with food and shelter by the swollen tissue. Eelworms and Fungi can also cause galls to be produced. 2. A painful swelling, particularly on a horse, or a sore caused by rubbing.

Gallon. A measure of volume or capacity, equivalent to 8 pints or 4.55 litres, used mainly for liquids but also as a dry measure for grain. The British Imperial gallon is equal to 277.3 cubic inches. The American gallon is equivalent to 0.83 British gallons.

Gambrel. 1. A bent stick once used to hang carcases. 2. A horse's Hock.

Game. In general terms, any animal particularly valued for sport or its meat.

Game Fowl. Breeds of Eowl descended from ancient fighting cocks, usually of low fertility, but carrying plentiful breast meat.

Gamma Garden. Place where the plants or insect pests have to be treated with gamma rays which are emitted from the source.

Gammon. The thigh and adjacent parts of a side of bacon.

Gander. An adult male goose.

Gang. 1. A number of employees working together on a particular job, harvesting crops, sometimes called gang labour and operating under a foreman (or ganger). 2. An area of pasture on which cattle are allowed to graze. 3. A set of tools used together, such as a set of Discs mounted on a single axle in a Disc Harrow.

Gantry. An annex to an oast normally with a slatted floor for ventilation, in which green hops are stored in pokes while the kilns are reloaded. Also called green stage.

Gapes. A disease mainly of young chickens and turkeys due to small worms in the windpipe, causing birds to gasp for breath or 'gape'.

Garget. 1. See Mastitis. 2. A swelling in the throat of cattle and pigs.

Gas Chromatograph. Instrument which is used in gas chromatography to detect the presence of valatile compounds.

Gas Chromatography. Separation technique which involves passage of gaseous moving phase through a column containing a fixed adsorbent phase.

Gastrectomy. The term used for the surgical removal of part or all of the stomach.

Gas Liquor. Ammonia.

Gas Store. 1. A sealed store in which fruit, particularly apples, are kept under controlled atmospheric conditions. Temperature is kept low, and the atmospheric levels of nitrogen and oxygen increased and decreased, respectively, in order to reduce cellular respiration so as to retard the rate of maturing and to preserve the fruit. 2. Sometimes applied to a gas-tight cold store in which fruit, particularly apples, are preserved by low temperature and accumulating carbon dioxide released by the fruit during respiration.

Gathering. A method of systematic ploughing in which the tractor and plough are turned to the right each time they come out or go into work, so that already ploughed land is circled in a clockwise direction. Usually carried out alternatively with casting when ploughing marked lands in a field.

Gauge Wheel (Plough). Small wheel which is needed for the control of the plough which is having short beam.

Gavage. The force feeding of poultry, usually with the aid of a stomach tube. Also called cramming.

Geiger Counter. By popular usage a Geiger Muller counter tube or such as tube together with its associated electronic equipment.

Gelatin. Protein which is extracted from skin, hair, bones, tendons etc. It is used in culture media for the determination of a specific proteolytic activity of microorganisms or for the preparation of a peptone.

Geld. To castrate (Castration), especially a horse.

Gelding. A castrated male horse.

Geletinase. Exoenzyme which degrades gelatin.

Genetic Effect of Radiation. Inheritable changes which are chiefly mutations and produced by the absorption of ionizing radiations. On the basis of present knowledge these effects have been purely additive, and there occurs no recovery.

Genetic Map. Arrangement of mutable sites on a chromosome which has been deduced from genetic recombination experiments.

Genetic Sterility. Type of male sterility which has been conditioned by nuclear genes in contrast to cytoplasmic sterility. It may be transmitted by either the male or female parent.

Genetic Stock Collection. Stock having known genes, translocations, inversions, additions or substitutions, and lines with resistance or tolerance to known races of pathogens, insects or weed pests.

Genetics. Refers to the science of heredity, including the study of its chemical foundation, its development, expression and its bearings on variation, selection, adaptation, evolution, breeding and the activities of man.

Genetic Engineering. A term applied to techniques, still being developed, which involve the separation of cells from collected 'early embryos', and nuclei transfer, by which multiple 'identical' offspring can ultimately be produced. These techniques are combined with those of Embryo Transfer to implant the 'engineered' embryos in host mothers. The term also embraces the techniques, also being developed, by which DNA altertion and Gene manipulation can hopefully enhance the genetic quality of stock and crops. Genetic engineering provides a powerful alternative to the time-consuming conventional methods of breeding (Cross-Breeding) and promises to revolutionise the agricultural industry by permitting scientists to raise greatly improved animals and crop plants (*i.e.*, larger, greater disease-resistance, improved food-conversion ratio, improved nutritional value, etc.).

Genotypic Selection. The selection of animals in stock breeding on account of their genotype, revealed by their performance following close study of their breeding behaviour and the results of progeny testing.

Gentles. A term for maggots.

Geoponics. The study of agriculture.

Georgic. Relating to agriculture.

Gerber Test. A test used by the Milk Marketing Board to determine the Butterfat content of milk. It involves the use of sulphuric acid and amyl alcohol to break down the proteins which surround the fat globules in milk, releasing the fat for measurement.

Germ. 1. A micro-organism causing disease, particularly Bacteria. 2. A loose term for seed or the nucleus of a seed.

Germicide. A chemical substance that kills germs.

Germination Cabinet. Type of seed germinator which is most commonly used in the seed testing laboratories.

Germination Test. In a germination test the seeds are kept under optimum environmental conditions of light and temperature for inducing germination. The conditions needed to meet legal standards are specified in the rules for testing the different kinds of seeds.

Germinative. Refers to the ability of the seed to sprout, grow and develop into normal plant.

Germinative Capacity. A term used in seed testing for the number of seeds in a sample which germinate in a given time.

Germinative Energy. The rate of germination of a seed.

Germplasm. Refers to an entire array of cultivars in a crop species, related wild species in the genus, and hybrid between the wild and cultivated species.

Germtrap. Simple device which is used to prevent the spread of some diseases.

Gestation Period. The length of time between conception and birth, during which a developing young animal is carried in its mother's womb. For cows about 283 days, for sheep 144-150 days, and for pigs 116 120 days.

Gibberellins. Plant growth stimulating chemicals which are able to induce a number of effects on plant *e.g.*, rapid stem growth, overcoming of dormancy, production of seedless fruit and other responses.

Gilchrist's Disease. Refers to the chronic infection of animals and man. It is characterized by suppurative and granulomatous lesions in lungs and other affected organs. The disease is quite common in dogs and only once in a horse. It is caused by fungus *Blastomyces dermatitides.*

Gilt. A young female pig not having produced a litter. Also called liet, yelt or yilt.

Ginning (Cotton). Process of separating lint from seed cotton in ginning factory.

Ginning Percentage (GP). Means weight of cotton lint obtained from seed cotton. It is usually expressed in terms of percentage of seed cotton or percentage of lint obtained from 100 unit of seed cotton by weight. It depends upon the cotton variety *eg.*, 40 *lb* or kg of lint obtained from 100 *lb* or kg of seed cotton by weight then GP is 40.

Glasshouse. A building constructed almost entirely of glass providing a sheltered environment, and warmer conditions than the open air, for the growing of flowers and vegetables. Artificial heating may also be provided. Also called a green-house.

Glass House Crops. Much of the glasshouse sector of the horticulture industry has been re-equipped since the mid-1960s. Widespread use is made of units with automatic control of heating and ventilation, semi-automatic control of watering, and carbon dioxide enrichment of the atmosphere. Tomatoes are the most important glasshouse crop and, together with lettuce and cumcumbers, represent some 95% of the total value of glasshouse vegetable output. The area under sweet peppers and celery has increased significantly.

Glat. A gap in a hedge. Repairing such gaps is known as glatting.

Glazed Fruit. When candied fruit is dipped for a moment in a boiling syrup to impart a glossy finish to it, drained and dried it is called glazed fruit.

Glean. To gather corn left by a reaper or by those hand-harvesting a grain crop.

Gluten. A mixture of insoluble proteins present in wheat (and other cereal grains) giving wheat flour its distinctive character. It is a sticky substance remaining after the strach has been extracted.

Goad. A pointed stick used to urge cattle to move faster.

Goaf. A term used in some parts for a stack of hay, straw, etc., in a barn.

Goatling. A female goat between one and two years of age.

Gobar Gas Plant. Also called bio gas plant. It is a simple plant for carrying out anaerobic digestion of crowdung and liter to produce a combustible gas which is useful for lighting and cooking purposes.

Gobbler. An adult male turkey. Also called stag turkey or turkey cock.

Go-down. A cutting in the bank of a stream or river allowing animals to get to the water to drink.

Gonad. Sexual gland, either ovary or testis or ovotestis, which produces the reproductive cells.

Gonadotropins. Hormones which are obtained from embryonic sex glands.

Goose. A larger bird of the duck family (Anatidae). Domestic geese are descended from the greylag goose and are mostly kept in small flocks under free-range, providing table birds. Some large commercial units exist. The term goose is usually applied to the adult female.

Core. A triangular piece of land.

Gosling. A young goose.

Gossypol. Phenolic pigment in cottonseed that is toxic to some animals.

Gout Fly. A small yellow and black fly related to the Frit Fly, the Larvae of which feed, in consecutive generations, on the shoots and the developing ears of cereals, particularly barley. The leaf sheaths become swollen (or 'gouty') and twisted. Also called the Ribbon-footed Corn Fly.

Grading Up. The establishing of a pedigree herd of cattle by mating pedigree bulls with a core of non-pedigree cows, and subsequently mating each generation of females with further pedigree bulls of the same breed, thereby increasing the percentage of pedigree 'blood' with each generation.

Graft. 1. A shoot of a desired plant variety (a scion) bearing a bud, joined on to the stem of another variety (the stock) with a weaker or stronger rooting system, as required, so as to develop into a whole plant, usually a tree or shrub, of the same variety as the scion. Grafting is a common method of propagation (Budding). Cherry varieties are often grafted on to wild cherry stocks. 2. A ditching spade with a long, narrow, concave blade. 3. A spade's depth. Also called spit.

Graft Incompatibility. Refers to the inability of parts of two different plants, when grafted together, to produce a successful union and of the resulting single plant to develop satisfactorily.

Graftage. The term used for the method of plant propagation in which, the limbs of two different plants are jointed together to form a new plant. There are many methods of grafting e.g., splice, cleft, saddle, side, and topworking etc.

Grain. 1. Cereal seed, either in bulk after harvesting or and individual seed. Also a growing cereal crop.

2. The smallest British Unit of weight, 1/7000 of a pound, 0.648 gm, equivalent to the average weight of a seed of corn.

Grain Drier. A machine used to dry grain after it has been harvested while too moist for safe storage with the possibility of fermentation occurring. There are three main types.

Batch Driers. The grain is held in batches in special containers and dried using a hot air flow, followed by cooling with a cold air flow.

Continuous Flow Driers. The grain is conveyed through the drier in a continuous flow in a shallow bed, either at rightangles to the flow of high temperature air (cross-flow type) or in the opposite direction (counter flow type).

Storage Driers. These are involving slow drying with large quantities of warm air until equilibrium is reached. They include bins or silos with perforated floors (in-bin-type) or buildings with flat concrete floors overlaid by ventilating airducts (on-floor type).

The rate of drying varies from 6% per day in silos to 6% per hour in small driers. The safe, long term storage of grain requires a moisture content of 14%.

Grain Drying Weight Loss. The following formulae are used for calculating grain weight loss on drying (Grain Drier). If X=weight loss; W_1=wet weight; W_2=dried weight; M_1=original moisture percentage; M_2=moisture percentage after drying; then :

(A) When weight f wet grain known : $X = \dfrac{W_1 x (M_1 - M_2)}{(100 - M_2)}$

(B) When after-drying weight only known : $X = \dfrac{W_2 x (M_1 - M_2)}{(100 - M_1)}$

Grain Lifters. Projecting devices fitted to the cutter-bar fingers of a combine harvester to lift cereals which have been flattened or laid-down by bad weather, so that the crop can be cut and gathered by the combine's reel.

Grain Moisture Content. The percentage moisture in harvested cereal grain. For long-term safe storage grain needs to be dried to 14% moisture content. Above this level there is a possibility that the grain will ferment and heat up, totally spoiling the grain if not well ventilated and occasionally turned.

Graminae. The botanical term for the grass family, flowering plants with long, narrow leaves and tubular stems, including pasture grasses, cereals, reeds (but not sedges), bamboo and sugarcane.

Granary. 1. A farm building for storing grain and other feeding stuffs (e.g. Cake Meal) and for processing them (i.e., grinding and mixing) before feeding to livestock. Also a grange. 2. A figurative term for a rich grain-growing area or region.

Grange. 1. A farmhouse with accompanying buildings and stables. Also a country house. 2. A Granary.

Grassland. Areas used for grazing stock, consisting mainly of grasses and clovers where cultivated, and also of mosses, lichens, heather, etc., where uncultivated or natural.

Grazing land. Any area of pasture, meadow or other grassland available for stock to Graze.

Grease. An inflammation of the skin of a horse's heels resulting in swelling, itching and the discharge of greasy pus.

Green Bacon. Bacon sold after curing with brine but before being smoked.

Green Feed. The term used for the mechanically harvested forage fed to animals while fresh and succulent. They constitute an excellent source of vitamin-A as they are rich in carotene.

Green Freeding. The feeding of a crop to animals in the 'green' or unripe condition.

Green Forages. A group of feedingstuff comprising the grasses and young cereals, some legumes (e.g., Clover, Sainfoin; Lucerne, Vetches, etc.) and certain Brassicas (e.g., Kale, Cabbage, Rape, etc.) green Maize, Beet tops, and Silage, etc. They characteristically have a higher protein content, particularly when very young, than succulent foods such as roots and tubers.

Green Harvesting. The cutting of a crop in the 'green' or unripe state.

Green Manure. A crop (e.g., Italian Ryegrass, Mustard, etc.) specifically grown for subsequent ploughing-in, in order to provide humus to the soil. Green manure crops are often planted in the autumn after a fallow period when no main rotation crop is to be planted, to prevent accumulated nitrates from being leached out (Leach) during the winter.

Green Manure Trampler. An implement which is used to trample and press the green manure crop in the fields.

Greenstuff. Green vegetables for the table, especially cabbages. Also called greens.

Grey-back. An old breed of geese with a mostly white body and grey wings and upper legs.

Grey Leaf. A deficiency disease of cereals, mainly of oats, but sometimes wheat and barley, due to inadequate manganese in the soil. Characterised by grey streaks on the leaves, causing many young plants to die and resulting in poor grain yield. Also called grey speck.

Greyface. A crossbred sheep derived from a Border Leicester ram and a Blackface ewe, produced mainly in Scotland. Greyface ewes are crossed predominantly with Suffolk rams, producing lambs with a high quality carcase.

Grice. A little big.

Grinding Corn. Cereal grain used for grinding in a Mill to produce flour or Meal.

Grip. 1. A drain, small ditch or furrow used to drain surface water off grassland. 2. A narrow opening made in the soil, using a spade, into which young plants or cuttings are placed. 3. A Dunging Channel in a Cowhouse.

Gripping Machine. A machine, normally drawn by a Crawler Tractor, similar to a large plough, used to cut surface drains or grips in boggy conditions in hill areas or river valleys, and usually fitted with wide tracks to prevent sinking. The machine normally has an arched frame supporting the cutting mechanism, which removes a continuous strip of soil leaving a U-shaped furrow.

Grist. Cereal grain which is to be or has been ground in a Mill. Also malt for brewing.

Grit. 1. Small fragments of various substances fed to poultry, viz. (*a*) soluble grift, such as limestone or oyster shell, provided to balance calcium lost by laying birds in producing egg-shells, and (*b*) insoluble grit, such as granite or fint, provided to assist the gizzard to grind food. 2. Small woody parts of a pear.

Grits. Coarsely ground Grain, particularly oats.

Groats. Grain, particularly oats, after removal of husks.

Grooming. The cleaning, brushing, combine and general tending of the coat, tail, legs and feet of certain animals, particularly horses and cattle, especially when kept indoors.

Gross Energy Value. The energy released as heat by a food when a given weight is completely oxidised, usually in a bomb calorimeter. Also known as energy of combustion.

Ground Chalk. Chalk, Ground Limestone.

Ground Frost. A temperature at ground level of less than 0°C or 32°F. Most plant tissues are not destroyed until the temperature falls well below the freezing point of water.

Ground-keeper. A potato missed during harvesting, which subsequently grows up again.

Ground Limestone. Any of the natural forms of limestone ground to produce a powder for use in liming. The powder acts to rectify soil acidity more quickly than limestone applied in lumps.

Ground Nut Cake, Ground Nut Meal. Cake or Meal derived from the tropical ground nut (*Arachis hypogea*) and available either in the decorticated or undecorticated forms. Used to feed dairy cows and fattening cattle as well as pigs, poultry and sheep.

Ground Rock Phosphate. Rock Phosphate.

Groundwater. Rain water which has percolated through to underground rock strata from the surface as distinct from draining to streams or rivers as runoff. Such water is the source of well and spring supplies.

Growing-off. A condition of hops in which the bines easily detach themselves from the rootstocks, due to Hop Canker.

Growth Promoter. A substance included in an animal feed (Feedingstuffs) which has the effect of increasing the efficiency of feed conversion (Food Conversion Ratio) and/or leads to improved daily liveweight gains. Examples included antibiotics such as flavomycin, virginiamycin and zinc bacitracin, and mineral copper sulphate.

Growth Ragulator. A chemical applied to a plant, which regulates or alters its rate of growth, either by stimulation or retardation. Often applied to control weeds in crops (*e.g.*, Hormone Weedkillers, Translocated Herbicides) when growth is severely retarded and death usually results. These chemicals act in a similar manner to auxins' the natural plant growth regulating substances or hormones.

Grub. A term applied in the general sense to insect larvae such as caterpillars and legless maggots, but sometimes used more specifically for the legged larvae of beetles.

Grub Out, Grup Up. To remove plants from the ground, particularly digging out the roots so that the land is cleaned. Especially applied to Orchards.

Grubber. 1. A hand implement for grubbing out weeds, etc., or for clearing and stirring up the soil, consisting of a long stick with a forked tip behind which is fitted a curled metal pivot. 2. A heavy cultivator.

Guano. A fertilizer rich in nitrogen and phosphorus, derived from the accumulated droppings of certain sea birds, found particularly on islands off the Peruvian coast. (Fish Guano).

Gypsum. ($CaSO_4 . 2H_2O$). Crystalline hydrated calcium sulphate. Often added to soil to correct saline conditions (*e.g.*, following seawater flooding), the calcium displacing the sodium (Cation Exchange). Superphosphate contains a certain amount of gypsum.

H

H-Building. Refers to a method of plant propagation by assexual means in which two longitudinal parallel cuts of equal length (3-4 cm long) are made in the bark of stock. A cross-cut is then made about midway between the ends of the two parallel longitudinal cuts. The two flaps of bark has been lifted and a bud removed from the scion with bark but without any wood has been inserted on the sock, covered by the flaps and tied.

H-Class. Nursery plots of one or two rows 2 or 3 m long and 40 cm apart used for testing new introduction, selections and hybrids under normal growing conditions.

Hack. A horse kept for hiring out, especially an old or poor one. Also an ordinary riding horse.

Hackles. Long, narrow shiny feathers on the neck of a cock.

Hackney. A breed of light draught horse, usually bay, brown or chestnut coloured.

Hag. A firm spot or higher spot in a marsh or bog. Also a place from which peat has been dug, often a hole filled with water.

Hagberg Test. A test used mainly by millers to determine the alpha amylase content of milling wheat. A suspension of 7 gm of ground wheat in 25 ml water, in a special tube, is stood in a boiling water bath and stirred. The suspension thickness and subsequently thins if the enzyme is present. After 60 second the stirrer is allowed to fall under its own weight and its time of fall is measured. A good milling wheat has a falling time of over 200 secs. Whereas with falling times below 100 secs. are not normally accepted for milling. Highly active wheats show falling time of less than 5 secs. Falling numbers are sometimes quoted. These include the total time in the water bath. Thus a falling time of 100 secs is equivalent to a falling number of 160 secs. (Farrand Test).

Hair Hygrometer. Instrument which is used for direct measurement of relative humidity.

Hake Bar. The part of a trailed plough which is coupled to the draught linkage behind the tractor.

Half-breed. Any animal of mixed breed. Specifically applied to a cross-bred. (Cross-Breeding) type of sheep resulting from a cheviot ewe crossed with a Border Leicester ram.

Half-Sib. Term used in population improvement. A Half-Sib family comprises progeny from a cross of bulk pollen into a selected parent—that is, one parent in the cross is identified (selected).

Half-standard. A fruit tree, the lowest branches of which are about 1.2 cm (4 ft.) above the ground. (Standard).

Halter. A rope, leather or canvas device fastened to the heads of cattle and horses for holding and leading them.

Ham. The thigh of a pig, usually cured by salting and drying in smoke.

Hammel. 1. A ram lamb. 2. A small shed.

Hammer Mill. A machine usually housed in a barn, which grinds Cereals and other Feedingstuffs. It consists of steel hammers which rotate at high speed in a casing and are surrounded by a perforated screen. When the grain is sufficiently reduced in size by the impact of the hammers it passes through the holes in the screen. Various mill sizes are available with differing screen mesh sizes, and throughout rates vary between 3 tonnes per hour for large powered mills and about 50 kg per hour for small automatic mills.

Hand. A measure of 4 inches (10.16 cm) used for describing the height of horses.

Hanging. The practice of hanging animal and bird carcases, particularly game, on hooks at room temperature for a period so that the meat condition improves.

Hanging Drop Technique. Technique for observing the presence of motile microorganisms in a drop of fluid.

Hank (Cotton). One hank is equal to 840 yards length of yarn. American cotton on an average is having 38 counts. It means 38 hank can be obtained from one proud of lint.

Hard Seeds. Clover seeds with very tough seed coats preventing Germination and which require scratching in order to do so.

Harden Off. 1. To allow a nursery plant grown in a Glasshouse to acclimatise to the more rigorous outside conditions by transferring first to a cloche or frame, and subsequently to the open air. 2. To gradually reduce the temperature in a brooder over a period of time until young chicks no longer need artificial heat to survive. Now-a-days temperature is usually reduced by 5.5°C (10°F) per week over the first four weeks of life, from about 32°C to about 15.5°C (90°F to 60°F).

Hardy. A term applied to plants and animals capable of surviving the cold during winter.

Hardwood Cutting Cuttings which are made from the past season's growth or olderwood that has shed its leaves and has got lignified e.g. grape, fig. etc.

Harrow. Secondary tillage implement which is used to cut the soil to a shallow depth for smoothening and pulverising the soil as well as to cut the weeds and to mix materials with the soil.

Harrow, Acme. Special type of harrow having curve knives.

Harrow Beam. That part of animal drawn harrow which is connecting the implement with the yoke.

Harrow Single Action Disc. Harrow with two gangs placed end, which throw the soil in opposite directions. The discs are arranged in such a way that right side gang throws the soil towards right and left side gang throws the soil towards left.

Harrow Spike Tooth. Harrow with peg shaped teeth of diamond cross-section attached to a rectangular frame. It is used to break the clods, stir the soil, uproot the weeds, level the ground, break the soil crust, and cover the seeds. Its principal use is to smoothen and level the soil directly after ploughing.

Harrow, Spring Tooth. Harrow with tough, flexible teeth, suitable to work in hard and stony soils. Spring tooth harrow is fitted with springs, having loops of eliptical shape. It gives a springing action in working condition. It is best suited for hard and stony ground. It is used in soil where obstruction like stones, roots and weeds are hidden below the ground surface. The type pulverises the soil and helps in killing weeds.

Harrow Tandem Disc. A disc harrow, comprising of four gangs or more in which each gang can be angled in opposite direction.

Harrow, Triangular. Spike tooth harrow with a triangular frame. The frame is made of wood and pointed spikes are fitted in the frame. The teeth of the spike are fixed and not adjustable.

Harrow, Zigzag. Spike tooth harrow with a zigzag frame and teeth attached at the junction of the frame members.

Harrowing. 1. Working the soil with the help of a wooden implement harrow. 2. Secondary tillage operation which pulverises and smoothens the soil.

Harvest. The time when ripe or mature crops are cut (eg., cereals), lifted (e.g., root crops) or picked (e.g., fruit, hops) and gathered in, the whole process being called harvesting (sometimes called ingathering). Sometimes applied loosely to the yield of a particular crop. Most arable crops now-a-days are harvested by machine, e.g., Combine Harvester, Potato Lifter Sugar Beet Harvester.

Harvest Index. The ratio of grain weight to total plant weight in a cereal crop.

Harvest Year. The first, second, third, etc., year following the Seeding Year of a Ley.

Hassock. A tuft or tussock of tightly packed grasses, reeds or rushes, etc.

Hatchel 1. A Haycock. 2. A comb for Flax or hemp.

Hatchery. A term applied (*a*) to either a cabinet or a walk-in type of incubator, or (*b*) to premises specialising in the hatching of eggs and production of Day-Old-Chicks, as distinct from the breeding or rearing poultry.

Hatching. The act of a young bird breaking out of an egg, aided by its egg tooth. A hen will normally sit on her eggs and incubate them to bring them to the point of hatching. Artificial incubation is used in commercial poultry rearing. A hatch is also a brood of chicks.

Haugh. A riverside Meadow or flat alluvial (Alluvium) land in a river valley, particularly in Scotland.

Haulm. The stems and leaves of corn, peas, beans, potatoes, etc, especially after harvesting. Also called halm.

Haulm Killer. A chemical applied to a potato crop to kill the above ground foliage to make harvesting easier.

Hay. A term mainly applied to grasses, but also to legumes, some herbs, and occasionally Cereals, which have been cut and dried and conserved for Fodder. Hay may be classified into two main types, viz., (*a*) seeds hay (sometimes called clover hay), usually derived from a 1 to 2 year old Ley, at one time mainly from a Red Clover-Ryegrass mixture, and (*b*) meadow hay, derived from permanent Grassland or a long ley. Normally stored in Bale form in a barn nowadays.

Hay Bale Stack. Bales of Hay built into a stack for drying and storage, often under a Dutch Barn, and usually on a base of rough timbers to raise the bales off the ground. Air spaces are left between the bales to allow moisture to escape and to allow lower bales to expand under the weight of higher ones. When chemical additives which prevent mould growth are introduced during baling, and hay is fully field-dried, bales may be more compactly stacked. Small temporary stacks are usually left in the field for a period to ensure full drying before barn stacking. Field stacks are occasionally thatched with hay or reeds, but more commonly a plastic sheet is used to keep out rain.

Hay Quality. Treatment and weather conditions during Haymaking can considerably affect Hay quantity and feeding value. Rapid drying is best, reducing cell Respiration and Carbohydrate loss by oxidation. Considerable Carotene loss may occur due to excessive bleaching by the sun if the hay is left in the swath too long.

Protein and ash are more concentrated in leaves and are better conserved in hay harvested whilst still in leafy condition. Rain and dew can leach soluble nutrients from swaths, and moulding, rotting and heating can occur in hay stored in too moist condition (i.e., above 25%).

Hay Rack. A metal or wooden open frame in which hay is placed and from which livestock may take what they require.

Haycock. A cone-shaped heap of loose hay drying in a field. Now usually only found on hill farms and smallholdings where the use of balers is not practical. Also called hatchel or pike.

Hayknife. A large broad-bladed knife with a cross-set handle at one end, used to cut Hay from a Haystack, or Silage from a clamp.

Haylage. A foddrer Crop wilted in the field, after cutting, to between 50-60% moisture content and then chopped and blown into an airtight Silo. Carbon dioxide released by cell Respiration inhibits bacterial activity and rotting. Drilage is almost identical except that wilting is only allowed to between 60% and 70% moisture content.

Hayloader. A trailed implement similar to a small elevator used to pick up loose Hay from a Swath and deposit it onto a trailer for carting to Haystack. Now seldom used except on some hill or small farms where the use of a baler is not practical.

Haymaking. The conserving of Hay for use as Fodder, involving drying and the maintenance of nutritive content. (Hay Quality). The moisture content of hay must be reduced from about 80% when freshly cut (Mower), to below 20% to prevent moulding rotting, heating, and possible combustion.

Hay-rake. A hand-rake with a long handle and normally with wooden prongs, used to gather hay.

Haysel. The haymaking season.

Hay stack. A storage stack of loose Hay common before the introduction of the Baler and bale storage. Now usually only seen in hill areas and on small farms where topography precludes the use of a baler. Plastic sheets, as protection against rain, have now replaced thatch roofs on such stacks. Some stacks of baled hay are referred to as haystacks. Also called hayrick.

Hay sweep. An implement, either horse-propelled or fitted to the front of a tractor (and sometimes to an old car), once used commonly to pick up hay from Swaths or Haycocks and to transport it to a stack. It consisted of a series of wooden prongs about 3 m (10 ft.) long and spaced at 30 cm (1 ft.) intervals in a frame. Often used in conjunction with an elevator or stationary baler. Nowadays most hay is collected using a pick-up baler.

Hayward. A person in the middle ages who had responsibility for fence and hedge repairs and preventing stock breaking through.

Haywire. Twine, normally polypropylene nowadays, used in a baler to tie bales. Now usually called baler twine.

Hazel. A common shrub coppiced in oakwoods and scrub, and in certain areas used for making hurdles.

Head Corn. The largest grains in a cereal sample as distinct from the smallest grains, or tail corn.

Headland, Headrig. The strip along the border of a field where a plough is turned during ploughing, and which is itself ploughed when the rest of the field is completed. Also called for racre.

Heart. A term applied to fertile soil capable of producing good crops (in good heart) or infertile soil in poor condition (in poor heart).

Heart Rot. 1. The decay of the heartwood of a tree as a result of fungal disease. 2. A disease of beet due to boron deficiency, resulting in the death and browning of the root centre and death of young leaves. Similar to Brown Heart.

Heat Treated Milk. Milk subjected to heating, to destroy pathogenic bacteria and those organisms which cause souring. The method of heat treatment differs for the various designations of milk.

Heath. An area of open, uncultivated, barren country, with poor acid soil, often sandy or gravelly, and with a characteristic vegetation cover of low shrubs, dominated usually by heather and other ericaeous plants.

Heavy Land, Heavy Soil. Land or soil with a high clay content, a high drawbar pull, and harder to cultivate than light soil, and thus sometimes called man's land.

Hectare. A metric unit of land measure. 100 acres of 10,000 sq. metres. Equivalent to 2.4711 Acres.

Hedge. 1. A close row of shrubs, bushes or small trees forming a fence or field boundary, and often planted or maintained so as to be stock-proof. Hawthorn is commonly used. Hedges last much longer than wire fences but are much more expensive. 2. To plant, maintain, trim or layer a hedge.

Hedge Cutter, Hedge Trimmer. A tractor-mounted machine used to trim hedges. There are three main types, viz. (*a*) flail type (Flail Hedger), (*b*) circular-saw type, with a heavy saw blade mounted on an articulated arm driven hydraulically by p.t.o., and (*c*) reciprocating cutter-bar type, also usually operated by p.t.o. There are also various hand-held machines for light or awkward work.

Heel. Back part of the food which is composed of one single bone, the calcaneus (as calcis). The heel pad is made of fat and skin which bears the weight.

Heel In. To store young plants temporarily prior to planting by keeping them in a trench and covering the root or rooting portions with soil.

Height (Tree). Means the vertical distance between the ground level and the extreme top of a tree. On slopes it is generally measured on the upper side of the tree.

Heifer. A term usually applied to a female cow over 1 year old which has not borne two calves. A maiden heifer is one that is still virgin. An in-calf heifer is a pregnant one.

Heifer Calf. A female cow less than 1 year old. Also called quey calf or cow calf.

Helme. A shelter for farm implements.

Helminth. Worm parasite.

Helophyte. Marsh plant growing in muddy, non-saline sites.

Hematic. Relating to blood.

Hematidrosis or Sweating of Blood. Due to the mixing of blood with the sweat because of some circulatory disease.

Hen. A female Fowl over 18 months of age, having completed the first Laying Period. Usually applied to domestic poultry. Also any female bird.

Haptachlor. An insecticide of the chlorinated hydrocarbon type, similar to and sometimes incorporated in chlorane, used for seed dressing.

Herb. A vascular plant which is distinguished from a tree or shurbs by having a non-woody stem. It is often used in medicine or for providing scent or flavouring.

Herbage. Herbaceous vegetation. It is generally applied to grassland species.

Herbicide. A pesticide which kills weeds. The action of a herbicide may be either selective, killing the weeds only and leaving the crop unharmed, or non-selective (or total), in which all vegetation is killed. The latter type are usually applied to fields before planting. Herbicides are normally applied either as foliar sprays direct to the soil where they have a residual effect. Applications may be before crop sowing (pre-sowing), after sowing but before crop emergence (pre-emergence), or after crop emergence (post-emergence).

Herbivore. An animal which eats grass and other herbage (e.g., cattle, sheep, rabbits, etc.) as distinct from a flesh-eating carnivore or an Omnivore.

Herd. A group of animals kept together for management particularly cattle. 2. A person who looks after a herd or flock, e.g., cowherd, shepherd, etc. Also an abbreviated form of shepherd.

Herd Improvement Test. Kind of production test which is carried out for dairy cows in which the whole herd, rather than any individual animal, is kept under test for the whole year.

Herd Test. Refers to a semi-official test for milk production of the whole herd in which all milk-producting animals are taken into consideration.

Hereford. A heavy and hardy breed of cattle, characteristically deep red with white markings on the head, back, chest, legs and tail tassel. A relatively early maturing breed, with short, strong legs, and a short, broad head. Sometimes polled now-a-days. A dominant breed throughout the world, known for it grazing ability.

Hesperidin. Crystalline substance which is found in orange and other fruits. Chemically it is a rhammo-glucoside of heoperetin, a flavanone closely related to flavone.

Heteroauxin. Growth-regulating substance such as β-indoleacetic acid.

Heterophytes Antibody. All the non-green plants which are not unable to prepare their own carbohydrate food but draw it from different sources. They are either parasites or saprophytes.

Heterosis. Literally means a condition which is different from parents, Genetically it is the increased vigour growth, yield, or function of a hybrid over the parents that has resulted from the crossing of genetically unlike organisms.

Heterotrophic Cells. Cells that need complex nutrient molecules like amino acids, glucose etc. from which to obtain energy and to build their own macromolecules.

Heterozygous. A genetical term for a plant or animal having a dominant gene for a particular characteristic (e.g., hair colour) derived from one parent and a recessive gene from the other, as distinct from two identical genes (when termed Homozygous).

Hexagonal System of Planting. Refers to planting fruit trees in an orchard at each corner of an equilateral triangle, by which 6 trees form a hexagon, the 7th tree being planted in the centre. This system has been found to be useful where land is expensive and very fertile with a good supply of water.

Hidden Hunger. Expression which is used to designate a condition in plants and animals brought about through the lack of certain essential elements in feed or in the soil and is not easilv detected by outward appearance.

High Farming. Farming to a high standard with good management and producing good yields, as distinct from low farming in which yields are low, due to poor management or infertile land.

High Pressure High Volume Sprayer. Hydraulic sprayer which is used for spraying tall trees and dense crops, similar to multipurpose which is multi-cylinder, plunger type, larger tank-size, can cover 10 to 20 acres a day.

High Temperature Short Time Process (HTST). A treatment in which heating of milk is done at a temperature of 160^0 to 162^0 for 12 to 20 seconds.

Hill Dropping. The method is used for sowing in which seeds are dropped at fixed spacing and not in a continuous stream. Thus the spacing between plant to plant in a raw remains constant.

Hill Placement. Application of fertilizer either in bands or in local areas near the plants like cotton, cabbage, tomato etc.

Hinge. A term applied to the uncut soil left by the share of wing of a plough which does not cut a full furrow width.

Hinny. The progeny of a stallion and female Ass; hardy, but sterile.

Hipped Cattle. Those injured by the horns of other cattle.

Hirsel. A group of heft of sheep. Also the area over which such a heft is grazing, often under the charge of a shepherd. Sometimes called a herding.

Hitch. The mechanism by which trailers and trailed implements are connected to a tractor for towing. The two main types are the Drawbar and the Pick-Up Hitch.

Hock. The projecting middle joint, of elbow-like appearance, on the hind legs of animals.

Hoe. A trailed implement used to cultivate the soil between row crops to control weeds, consisting of horizontal blades drawn just below the soil surface. Various types exist including Tractor Hoes, Steerage Hoes, and front-or mid-mounted hoes which are rigidly connected to the tractor and are raised or lowered hydraulically, and can be operated by the tractor driver. Hand does comprising a thin blade on a long handle are used by gardeners to loosen soil and to chop and removed weeds.

Hog. A castrated male pig. Also called a bar or barrow pig.

Hog Cholera. Acute, septicalmic, highly contagious and fatal disease or pigs. It is characterized by high fever, anorexia; severe leucopenia, haemorrhages in different parts of the body and it is after accompanied with digestive, nervous and pulmonary symptoms. It gets caused by a filterable virus with a particle size of about 35mμ. The affected animals eliminate the virus in large quantities in all their excretions and secretions.

Hog Flu. Acute,highly contagious viral disease of pigs which is characterized by sudden onset with fever, respiratory symptoms and marked prostration and is caused by a virus related to the human influenza virus.

Hogg, Hoggerel, Hogget. A male or female sheep between being weaned and being shorn for the first time. Sometimes spelt hog. Also called teg.

Hollow Land. Recently ploughed land which has not settled and compacted again.

Holm. Flat, fertile and adjacent to a river.

Holstein. Friesian cattle first imported into Canada in 1881 from Holland and bred there to produce a variety with dairy-like characters (*i.e.,*less flash) and giving high milk and butterfat yields (Milk Yields). Characteristically black and white in colour. Also called the Holstein-Friesian or Canadian Holstein.

Holt. 1. An Orchard. 2. A wooded hill.

Homogenised Milk. Milk made more digestible by breaking up the fat globules so that they are evenly distributed instead of becoming concentrated as cream.

Homozygous. A genetical term for a plant or animals having two identical genes for a particular characterisitic (e.g., hair colour), one genes derived from each parent. The genes may be both dominant or both recessive.

Honey. A thick, sweet fluid prepared by worker bees in a Beehive from nectar collected from flower and stored in the honeycomb (Comb) for feeding to the larvae. Different honey with characteristic flavours are produced from nectar derived from different plant species.

Honeydew Honey. Honey produced wholly or mainly from secretions of, or found on living parts of plants other than the flowers and which is light brown, greenish brown, black or of any intermediate colour.

Hoof and Horn Meal. A fertilizer, rich in protein, consisting of dried and ground animal hooves or horns (or a combination of both). It has a nitrogen content of 12-14% which is rapidly released when appplied to warm soils, normally before planting or sowing. Also called Hoof Meal, and sometimes Horn Meal.

Hoover. A type of potato digger with a chain elevator.

Hop. A climbing plant (*Humulus Iupulus*) of the mulberry family which is characterized by a long, rough twining stem and rough vine-like leaves (hop bine) and is bearing clusters of bitter, catkin-like fruits of cones (the hops) used for flavouring beer.

Hop-acre. A unit of hop production consisting of 1200 individual hop plants (Hills), and sometimes 1000 plants, depending on the planting distance.

Hop-dog. A hand tool with a wooden handle and serrated iron jaws used to remove poles (Hop) from the ground.

Hop Mildew. A fungal disease (*Sphaerotheca humuli*) of hops, causing white spots on leaves, and attacking the female flowers and preventing cone formation. Sometimes late attacks occur causing cones to become reddish coloured (red mould).

Hoppus Food. A unit of measurement of round timber equal to 1.273 cubic ft. The extra 0.273 ft compensates for surplus wood assumed lost during sawing to produce squared planks.

Hormone Weedkiller. A synthetic hormone applied to growing crops which acts selectively by causing distored growth and eventual death to weeds whilst leaving the crops unaffected, *e.g.*, 2,4-D, M.C.P.A. Usually called a growth regulator.

Hormones. Organic substances produced by plants and animals in minute quantities. Plant hormones (*e.g.*, Auxins) are involved in growth regulation. Animal hormones (*e.g.*, adrenalin) are usually secreted by various endocrine glands into the blood stream, and affect behaviour and a variety of body functions.

Horse Fly. A general name for the blood-sucking, two-winged tabanid flies *e.g.*, Breeze Fly, Cleg, Gadfly.

Horsefly Repellent. Horseflies in small dirty herds could be repelled by employing following mixture.

Content	*Parts w/w*
Water	2
Tar oil	1
Fish oil	1
Kerosene	1
Soap shavings	1
Pulverized Rosin	2
Water	0.5

The soap and rosin are boiled in water until dissolved. The remaining ingredients are then added and the whole mixture is boiled for 15-20 minutes. The mixture should be applied every 2-3 days.

Horticulture. In strict terms the scientific cultivation of fruit , vegetables, flowers and shrubs. Also used to describe the commercial production of such crops, some on general farms, but mostly on specialised holdings where soil, climate, skilled labour and irrigation can produce maximum yields of high-quality crops and where access to markets or good roads enable their sale at economical prices. Most horticultural enterprises have increasing output per unit area with the help of cultivation and environmental control, and the widespread use of machinery. On some farms the shortage and high cost of labour have led to the introduction of 'pick your own' harvesting for direct sales to consumers.

Host Indexing. Refers to a procedure which is used to determine whether a given organism (Plant or animal or microbe) is a carrier of a virus disease. Material is taken from that organism an transferred to another that will develop characteristic symptoms if affected by the virus disease in question.

Host Range. Various kinds of plants or animals which could be affected by a given pathogen.

Huller. A mobile machine which operates like a threshing Machine and is used to separate the seed from the head of clovers, etc.

Humectant. Substance that is able to absorb moisture. It is used for maintaining the water content of products like tobacco, glue, films, dentifixices, cosmetics, baking products, leather, soaps and textiles.

Humidity Mixing Ratio. Defined as the amount of water vapour in grams mixed with one kilogram of dry air. Common values are in the order of 4-30 gm per kg. For each temperature there is a corresponding saturated humidity mixing ratio.

Humification. The term used for the process of decomposition of organic matter leading to the formation of humus.

Humus. Decomposed and partly decomposed Organic Matter in the soil derived from plant and animal remains as a result mainly of bacterial action, giving a dark colour to the upper Soil Horizons. Humus has colloidal properties and plays an important role in Cation Exchange, and assists in giving Crumb structure to soil.

Humus-nucleus. A stable complex of Lignin and Protein derived from decaying plant remains, present in the soil.

Hungry Soil. One requiring plentiful supplies of fertilizer to produce good crops, due to lack of organic matter and relative infertility. (Sandy Soils).

Hurdle. A mobile wooden or metal frame used for temporary fencing, particulary for sheep.

Husbandry. The economic management of a farm.

Husk. 1. The dry, thin, outer covering of certain fruits and seeds, *e.g.*, the Glumes of cereal seeds. 2. A disease of cattle and sheep due to parasitic nematode worms, and typified by a husky cough and, when severe, difficult breathing.

Hybrid. The offspring of parents of different species, varieties or breeds of plants or animals. They may be fertile or sterile. The likelihood of sterility is increased the greater the difference between the Genotypes of the parents, due to the increase chance of imperfect Chromosome pairing.

Hybrid Vigour. Qualities in a Hybrid not present in either parent, *e.g.*, increased hardines, improved growth rate.

Hydrated Lime. Calcium Hydroxide.

Hydroponics. The growing of plants (*e.g.*, tomatoes) in nutrient solution without soil.

Hyena Disease. An irreversible wasting condition of the hind-quarters in dairy heifers. The disease has not yet been recognized in Britain but is well known in France and the Low Countries.

Hygiene. Science of health and its preservation.

Hygrograph. Instrument which is used for recording automatically and continuously the variations of the relative humidity of the atmosphere.

Hypervitaminosis. Undesirable effects which are produced by taking an excess of a vitamin concentrate or pure vitamin.

Hypodermic Injection. Administration of a liquid remedy by injection into the subcutaneous connective tissues.

Hypsometer. Instrument which measures altitude.

I

I-Budding. Refers to a method of plant propagation (asexual) in which the bud patch is cut just as a rectangular or square patch. Two transverse cuts are made through the bark of the stock. These are made to join at their centres by a single vertical cut to produce the shape of I. The two flaps of bark are then raised for inserting the but patch beneath them. This method is used if the bark of the stock is much thicker than that of the budstock.

ID_{50} (Infectious Dose, 50%). Dose (number of micro-organisms) of a given infectious agent which will be able to infect 50 per cent of the experimental animals in a test series under given conditions.

I_1, I_2, I_3. Symbols which are used to designate the first, second and third generations of inbreeding.

Ileum. Last portion of the small intestine which extends from the deodenum to the cecum.

Iliac. Pertaining to the ilium.

Ilium. Flank bone, which is one of the bones forming the pelvis.

Illegitimate Pollination. Self-pollination, which takes place, inspite of the flower appearing to be adapted for cross-pollination.

Imago. A sexually mature adult insect; the final stage of Metamorphosis.

Imhibitional Water. Moisture internally absorbed by the clayhumus complex in the soil. (Capillary Water, Gravitational Water).

Imhoff Tank. Two-story sewage-digestion tank where in action is predominantly be anaerobic microorganisms.

Immunisation. The inducing of resistance to an infectious disease or poison in an animal, normally by injecting a dose of vaccine or antitoxin.

Immunity. The ability of a plant or animal to resist an infectious disease or the effects of a poison.

Improved Dartmoor. A longwool breed of sheep developed by crossing old moorland Dartmoor sheep with the Devon Longwool breed. Found in localised areas only, mainly in semilow-land situations.

In-. A prefix used descriptively for a pregnant female of the species suggested by the added name of the appropriate young animal, which is being carried in the womb, *e.g.*, in-alf, in-lamb, in-pig, etc.

In Burr. A stage in the growth of hop bines when the female flowers are fully developed and the stigmas are proturding.

In Lay. A hen during an egg laying Period is said to be in lay.

In Milk. A cow during a lactation period is said to be in milk.

In Season. A female animal in the state ready for mating. Also called 'on heat'.

In Vitro. In glass. Applied to biological experiments which are performed in test tubes or other laboratory glassware.

In Vivo. Within the living organism; applied to laboratory testing of agents within living organism.

Inactivation. The destroy the activity of a substance *e.g.*, heat treatment (56°C for 30 minutes) of serum abolishes complement activity in the serum.

Inarching. The grafting of a growing branch to a stock, without separating it from its parent stem. (Graft).

In-breeding. The mating of closely related animals (or plants), *e.g.*, parent with offspring, brother with sister, etc., to increase the number of individuals bearing specific highly desired characteristics. The practice tends to increase the number of Homozygous gene pairs, but genetically linked undesirable characteristics are also accumulated and progeny with such characteristics are usually culled. (See Breed in, Cross-Breeding, Line Breeding, Out-Crossing).

Incubation. The hatching of eggs, either naturally by a hen sitting on them, or artificially by keeping them in an artificially heated Incubator. (Incubation Period).

Incubation Period. 1. The period of time required for a newly laid egg to develop to the stage when the young bird is ready to hatch out. For chicken eggs the period is 21 days. 2. The period between an animal becoming infected by disease-causing germs (*e.g.*, Bacteria) and the symptoms developing, during which germ number increase dramatically.

Incubator. An apparatus used to hatch eggs by providing artificial heat, usually maintained at 37.2°C (99°F) and reduced for the last 2 days of the Incubation Period to about 36.1°C (97°F). Humidity is maintained at about 60%. Types vary from small flat ones holding 50-100 eggs, used for a small breeding flock, to large cabinet types holding up to 40,000 eggs in trays or specially designed walk-in-rooms holding up to 80,000 eggs, normally used at a Hatchery. Trays are frequently tilted to turn the eggs before they are transferred for the final 2 days of incubation to a hatching section.

Indian Game. A heavy table breed of poultry, variously coloured, including dark, white, white laced, or mottled (red, white and black), with yellow legs, skin and flesh, and producing tinted eggs. Known for its meat quality and used in crossing. Also called Cornish.

Infectious Bovine Rhinotrachoitis. An acutely contagious viral disease affecting mainly the upper respiratory tract of cattle. Symptoms include runny eyes and nose, salivation, high temperature and loss of appetite together with a drop in milk yield in dairy cows. Fatalities are not common. The disease is more severe in store cattle.

Infectious Bronchitis. Acute, highly contagious viral disease of chicken which is characterized by respiratory symptoms like gasping and drastic reduction in egg production. It is caused by filterable virus having a particle size of 70-106 mμ.

Infectious Disease. Those diseases capable of transmission from animals suffering from the disease-free animals.

Infectious Equine Encephalomyelitides. Characterized by symptoms of incoordination of movement and paraplegia. It has been caused by at least four distinct viruses which differ from each other in antigenic characters, mode of spread, verulence etc.

Infectious Laryngotrachitis. Acute, highly contagious viral disease of chickens, characterized by gasping respiration rates, sneezing and coughing which may be accompanied by expulsion of blood stained mucus from the nostrils. It has been caused by a filterable virus which can pass through a pore diameter of 0.80μ.

Infectious Necrotic Hepatitis. Acute infectious toxaemic disease which affects affecting sheep and cattle. It is characterized by short course, blackish discolouration of the skin caused by subcutaneous uenous congestion and necrotic foci in liver. It is caused by localized bacterial infection of liver by *Clostridium oedematiens.*

Infectious Ovine Encephalomyelitis. Acture viral disease of sheep which is occasionally affecting horse, cattle pigs and man, characterized by fever, peculiar louping gait, convulsions and paralysis. It is caused by Russian tick born complex of the group B arbor viruses.

Infertile Soil. A soil which is deficient in plant nutrients and requires plentiful applications of fertilizer to produce good crops.

Infield. An old Scottish term for a field near the farm buildings under continuous arable cultivation and constantly manured. (Outfield).

Injurious Weeds. Certain harmful weeds (*i.e.,* spear thistle, creeping thistle, curled dock, broad leaved dock, and ragwort) which the Ministry of Agriculture, Fisheries and Food may require an occupier to prevent from spreading, by written notice issued under the Weeds Act 1959. Also called Scheduled Weeds.

Inlay Approach Graft. Type of approach grafting to be used where the bark of the stock plant is considerably thicker than that of the scion plant.

Inoculation. Artificial introduction of micro-organisms on or into a medium, living system (*e.g.,* animal, tissue culture). Inoculation is carried out with an aseptic technique.

Inorganic Insecticide. Insecticide of inorganic origin like arsenical or fluorine compounds; lead or calcium arsenate and paris green examples.

Insect(s). A class of Arthropod with bodies divided into a head, thorax and abdomen, the head bearing a pair of feelers or antennae, the thorax three pairs of legs and wings. Adulthood is preceded by three stages: egg, larva and Pupa. Insects include Flies, fleas, Beetles, Aphids, Bees, etc.

Insecticide. A pesticide which kills Insects. Most of those in farm use today are synthetic organic compounds and include Chlorinated Hydrocarbons, organo-phosphorous compounds and Carbamates. They are available in granular, liquid and powder forms and may be applied in various ways.

Inseminate. 1. To sow seed. 2. To artificially transfer Pollen to the Stigma of a plant.

Inside Bats. Poles between the Straining Poles in a Hop garden which support the Top Wire.

Insectary. Also known as insectarium. A laboratory in which insects are bred in large numbers.

Insecticide Solvent. Substances in which insecticide in the form of crystal or powder is dissolved so that it might be used as spray; water commonly used; acetone, ethyl alcohol, benzene, xylene, linseed oil, groundnut oil, cotton seed oil, turpentine, kerosene, petroleum and other solvents; methylnaphthalenes superior solvents.

Intensitometer. A device which is used for determining relative X-ray intensities during radiography in order to control exposure time.

Intensive Farming. A method of farming in which the aim is to produce the maximum number of crops per year, of high yield, from the amount of land available and to maintain a high Stocking Rate for livestock. (Extensive Farming).

Intensive Livestock Production. The keeping of certain livestock (*e.g.*, beef, pigs, poultry, etc.) mainly indoors, often in relatively large numbers, with the aim of maximising efficiency by reducing per capita costs (*e.g.*, labour, equipment, feed, etc.) and the area required. A wide variety of management system exist. Also called factory farming. (Battery Hens, Piggeries).

Intensive Protection. Protection of crops from pest damage to the extent that such damage is negligible and does not impede precise measurement of research parameters.

Inter Body Clearance. The diagonal distance between the bodies of a plough.

Inter-Cropping. The growing of two or more totally different species together in the same field. A common practice in developing countries which provides an insurance if one crop fails.

Inter Crops. The crops which are raised in an orchard or other widely spaced crops for increasing the income from the same piece of land by short duration crops like vegetables, pulses, papaya etc.

Interbreeding. Experimental hybridization of different species of varieties.

Inulin. Starch found in tubers and roots of artichokes, dahlias and dendelions. It is a fructosan as it is hydrolyzable to fructose. It readily dissolves in warm water and does not give any colouration with iodine.

Inverse Nitrogen: Yield Concept. According to Wilcox, the yield of a crop has been found to be inversely proportional to its nitrogen content *i.e.*

$$y = \frac{k}{n}$$

Where y represents the yield

n represents the percentage of nitrogen in the crop and

k is a constant, the value for which was found by him to be 318 1b/Acre.

Iodophor. An iodine containing liquid used for teat dipping to prevent the spread of Mastitis in dairy cows.

Iron Bacteria. Bacteria in the soil which secrete hydrated ferric oxide in badly drained soils whose water contains ferrous bicarbonate in solution a condition made possible only if calcium carbonate is absent.

Irradiated Milk. Milk in which the vitamin D content is increased by irradiation by ultra violet rays.

Irradiation Breeding. Creation of mutations by subjecting the plants or seeds to radiations and their utilization for crop improvement.

Iron. (Fe). A Trace Element which, if deficient in the soil (particularly calcareous soils), can cause Deficiency Disease (Chlororis) especially in fruit crops. Also essential for Heamoglobin formation (Deficiency causes anaemia in animals) and for a variety of metabolic (Metabolism) needs in both plants and animals.

Irrigation. The application of water to soil to provide an adequate supply for crop needs, to increase crop yields or to aid their establishment. Mainly used for grass, vegetables, potatoes and sugar-beet in India. Water may be supplied by the mains, by abstraction under licence (Abstraction Licence), from a watercourse, or from on-farm reservoirs. Application is measured in inches per acre (Acre-Inch) or milimetres per hectare, the amount and frequency being related to various considerations including Soil Moisture Deficit, soil type (Light Soils contain less Readily Available Moisture than Heavy Soils), rainfall pattern and Evapotranspiration.

Irrigation Efficiency. The ratio of the amount of water used by irrigated crops to the amount actually supplied.

Irrigation Use of Water. Use of water primarily for the purpose of irrigation crops.

Irrigation Water. Water which is artificially applied in the process of irrigation. It does not include precipitation.

Isotopic Tracer. An isotope of an element, (radioactive or stable) a small amount of which may be incorporated into a sample material (the carrier) so as to follow the course of that element through a chemical, biological, or physical process, and also follow the larger sample.

Itching. Skin irritation which may occur because of chronic bowel irritation, chronic kidey diseases or a prolonged feeding of a restricted diet.

J

Jacobsen Germinator. Special device which is being used for seed testing wherein water is supplied to a blotting paper substrative by using wicks.

Jam. Concentrated fruit pulp having a fairly heavy body form rich in natural fruit flavour. Pectin in the fruit imparts it a good set and high sugar concentration (more than 68%) and facilitates its preservation. It is prepared by boiling the fruit pulp and juice with sufficient quantity of sugar to a reasonable consistency to hold the fruit tissues in position. A good jam must be having bright colour, rich flavour typical of a fruit, should not be syrupy or stiff. It should be free from crystallization of sugar.

Japanese B Encephalitis. Virus infection which is affecting not only anim is but also human beings. The disease is seasonal and the outbreaks generally remain confined to the warmer months. The disease causes fever, loss of appetite, lethargy, sleepy appearance and jaundice.

Jaundice. Yellowish or orange discolouration of the tissues and secretions. It is especially visible in eyes, skin and mucous membranes. It is caused due to circulation of bile pigment in the blood.

Jellies. Semi-solid product which is prepared by cooking essentially a clear fruit extract and sugar. In jelly making pectin forms the most essential constituent. A perfect jelly should be transparent, attractive in colour, give strong flavour of the fruit and firm enough to retain a sharp edge when cut but quivers when pressed. The fruits rich in pectin and acid like sour apple, grapes, guava, lemon, orange, loquat etc. are best for making jelly.

Jersey. A relatively small, early maturing breed of dairy cattle imported from Jersey. Various coat colours exist from light fawn to darkish red, almost black, sometimes with white markings. Characterized by a lightish coloured ring encircling the black muzzle, with a small head and dished face. Leanly built with a large well-developed udder. The average annual milk yield (Milk Yields) in 1980/81 was 3879 kg and the rich creamy milk contains 5.22% butterfat and is much used for butter making.

Jersian. A hybrid type of beef cattle derived from a Jersey bull crossed with a Friesian cow. Also called an F-J hybrid.

Johne's Disease. An infections bacterial disease (*Mycobacterium paratuberculosis*), mainly of cattle (other animals including sheep are sometimes, affected), characterized by severe inflammation of the intestines continuous diarrhoea, weakness and rapid loss of condition leading to emanciation.

Jointer. A term for a skim coulter.

Joint-ill, Joint-felon. A disease of young livestock caused by various Bacteria entering the body via the unhardened navel, and characterized by abscess formation at the naval and swellings in some limb joints. Also called navel-ill.

Jungle. (1) Low or thin forest (2) Any impenetrable thick or tangled mass of vegetation (3) Any dense intermingled growth.

Juvenile Form. Young plant that is having leaves and other features different from those of a mature plant of the same species.

Juvenile Hormone. Hormone which is secreted in insects by small cervical glands called *Corpora Allata.* Its presence in the body of the insects results in the conservation of larval character and the inhibition of metamorphosis.

Juvenile Stage. Special stage in the life history of some algae from which the ordinary plant is developed as an outgrowth.

Juvenility. In some plants, the differences in the internal stages of young seedlings from that of mature plant have been manifested by striking differences in leaf shape or greater thorniness in younger plants. This immature condition is referred to as juvenility. It is important in several aspects of plant propagation.

K

Kahn Test. Flocculation test which is used for diagnosis of syphilis in which cardiolipin (a non-treponemal) antigen is used. Such tests are also called standard serological tests (SSL).

Kainit. A low-grade fertilizer containing 12.30% potash and about 50% common salt used mainly on sugar beet. Sometimes also containing about 10% magnesium.

Kale. A crop of the Cabbage group characterised by a strong tall stem and open curled leaves, grown as a Fodder Crop. Fed either in the field or after cutting, or made into Silage. There are several types, viz, Hungry Gap Kale, Marrowstem Kale, Rape Cale, and Thousand-Headed Kale.

Ked. A brown, flat, wingless, hairy, blood-sucking fly (*Melopoagus ovinus*), parasitic on sheep, and causing constant irritation. Also called cade, kade, kaid, seep louse and sheep tick.

Keep. Grass or other Fodder Crops on which livestock are grazed.

Kelps. The largest of the brown algae which usually consist of expanded leaf-like blades connected by a stalk to a holdfast.

Kemp. Coarse, brittle, white fibres, shed during the growth of a Fleece, which are difficult to dye.

Keratin. A tough, sulphur-containing, fibrous protein. The major constituent of the outer layer of skin and horn.

Ketchup. Strained concentrated extract of particular fruit vegetable which is having spiced vinegar.

Kibbled. Broken into coarse fragments, *e.g.*, kibbled maize, kibbled lime.

Kid. A young goat less than 1 year old.

Kieserite. ($MgSO_4$ H_2O). Magnesium sulphate, a greyish white crystalline powder containing about 16% magnesium, used as a fertilizer. Regarded as a concentrated form of Epsom Salts, having less water for crystallisation.

Killing Out Percentage. The weight of a dressed carcase as a percentage of the Liveweight of the animal. (Deadweight). An indication of the amount of meat on a carcase after the removal of the head (except in pigs), limbs, hide, blood and offal. For beef cattle the average is 54-57%.

Kip. The hide of a young animal.

Knacker. An animal slaughterer.

Kitchen Gardening. Practice of growing vegetables in home gardens or compound gardens on a small scale.

Kline Test. Microscopic flocculation test for the diagnosis of syphilis.

Knives. Reciprocating sections which slide to and fro across the flat faces of the Fingers on the Cutter Bar of a Mower, Binder or Combine Harvester.

Knock Down. Percentage of immediate kill achieved by the application of an insecticide.

Knot. Portion of the branch which is embedded in the wood by the natural growth of the trees. The knot is 'loose' or 'tight' depending on whether the branch was dead or living at the time it was embedded.

Knotter. A mechanism on a Baler or Binder which knots the twine binding the Sheaves. Also called knotter bill.

Kohl Rabi. A fodder crop of the Cabbage group with a much swollen, turnip-shaped, scarred stem, from which grow long-stemmed leaves.

Ksloes. A mainly black, shaggy haired, small breed of cattle once found on the Hebridean islands and which contributed, with mainland cattle, to the development of Highland cattle. Sometimes used to mean Highland cattle.

L

LD_{50}. (Lethal Dose 50%) Dose of a given lethal agent (say number of pathogenic microorganisms) which on given to each of the test animals, will be able to kill 50 per cent animals in a test series.

Labelled Compound. A compound having labelled molecules. By observations of radioactivity or isotopic composition, this compound or its fragments may be followed through physical, chemical, or biological processes.

Lac. Resin which is secreted by the glands of the body of lac insect. Only resin from animal source. It is having peculiar properties, making it a versatile natural resin, lending itself to diverse applications in industry. There is 70-80 per cent of world production in India, over 90 per cent of lac produced exported.

Lac Insect (*Laccifer Lacca* Kerr). Insect belonging to group of occids (scale insect and mealy bugs). It is having a large number of small lac glands all over the body producing the lac resin; quantity produced small; needs secretion of 1,50,000 insect to make up one *lb* of shellac. It thrives on sap of host plants.

Lacciferidae. Family of scale insects including the Lac Insect *Laccifer lacca*. They secrete a resinous substance from which shellac is prepared.

Lactation. The period during which a female animal is secreting milk from the Mammary Glands. In cows and goats kept for milk production the period is greatly increased by regular milking, without which secretion ceases. The optimum lacation for dairy cows is 10 months (305 days) from calving to when twice dairy milking ceases, thus allowing a 2 month rest (dry period) prior to calving again. Milk yield during lactation varies with a peak at about 10-12 weeks after calving. Various factors affect the yield and quality of milk including breed, diet management and length of dry period.

Lactic Acid. An organic acid formed by bacterial action on lactose in milk, causing souring.

Lactoflavin. Term sometimes used for riboflavin or vitamin B_2. It is a watersoluble, yellow pigment having a greenish fluorescence. It is found in milk whey, egg yolk, egg white, leaves of plants and animal tissues.

Lactometer. Modified form of common hydrometer which is used for testing the purity of milk. The scale on the stem has been marked M, 3, 2, 1 and W from the bottom upwards. Milk being heavier than water, the instrument sinks less in milk than in water. In pure water it sinks to the mark W but in pure milk it sinks up to the mark M.

Lactory. A Milking Parlour.

Lactose. A disaccharide sugar (Carbohydrate) found in milk. A compound of glucose and galactose molecules, into which it is split by the enzyme lactase contained in Intestinal Juice.

Laid Crop. One, particularly a cereal crop, flattened in the field by the weather conditions, *e.g.*, wind or rain, or because the stem is not strong enough to support the grain vertically. Also called a lodged crop.

Laid Hedge. A hedge that has been layered.

Laid Up Field. One reserved for a particular purpose, *e.g.*, kept ungrazed for Hay production.

Lairage. A place where livestock are housed, particularly at markets and docks while awaiting slaughter or export.

Lamb. A young sheep under 6 months old, or the meat derived from it. Also to give birth to lambs.

Lambing Percentage. Inflammation of the lamina in an animal's hoof, particularly cattle and horses, thought to be an allergic reaction to overfeeding with barley, or to toxic products from afterbirth breakdown or bacteria.

Laminarin. (1) Polysaccharide storage product which is found in brown algae (2) Carbohydrate food reserve in brown algae.

Lammert's Cycle. Daily cycle of temperature changes taking place in beehives during winter. The temperature decreases to a minimum of 13°C. This stimulates the bees to greater muscular activity and the temperature increases rapidly to about 25°C. At this temperature the bees stop their activity and the temperature again decreases gradually to 13°C. In an average hive about 20 gms of sugar are used up during this cycle.

Land Drainage. 1. The construction of drains in or under a field to remove surplus water from the land to a ditch. Such drainage stabilises soil structure, improves aeration and root development, allows the soil to warm up more rapidly, promotes early crop growth, and lengthens the growing season for grass and arable crops. It also reduces the incidence of certain plant and animal diseases, and weed growth. 2. The removal of surplus water from land via ditches, streams and rivers.

Lands. Strips or divisions marked out in a field due to be ploughed by dividing ridges or rigs, each subsequently ploughed systematically (Casting, Gathering). The width of each land is determined according to tractor and plough size, *e.g.*, 40-50 m wide for a 4-furrow plough.

Landside. A long piece of metal fitted to the frog of a plough which bears against the furrow wall on the unploughed side and resists the force of the Mouldboard as it turns the furrow, thus providing lateral accuracy and stability. The landside sometimes includes a secondary spiring-loaded, wheel-like part which rolls and reduces wear.

Lanolin. Wool wax having greater purity than the ordinary wool grease of common degrees; the USP and cosmetic grades of lanolin have been highly purified, refined, bleached and deodorised.

Lanugo. Hairy covering on a foetus, begins to be shed before birth.

Laterals. Side shoots which develop from lateral buds on the branches or stems of trees or shrubs, as distinct from terminal shoots which develop from terminal Buds.

Latex. Milky, usually white, fluid emulsion which occurs in certain cells in some famous gum resins, fats, waxes, and often complex mixtures of other substances, frequently including poisonous compounds rubber, gutta-percha, chicle and balata are the chief commercial products obtained from latices.

Lauren's Half-Shade Polarimetre. Instrument which is used for finding the optical rotation of certain solution. When it is used for finding the optical rotation of sugar it is called a saccharimetre. If the specific rotation of sugar is known, the concentration of the sugar solution can be found out.

Lauritsen Electroscope. Rugged yet sensitive electroscope which is using a metalized quartz fiber as the sensitive element.

Lawn. Open ground about a house or other building which is grown to the grasses and maintained in good turf, especially for its aesthetic value.

Lawn Mover. Mechanical device which is used for cutting lawn grass to an even height.

Laxatives. Mildest and least active cathartic which is used to overcome constipation *e.g.,* white mineral oil, sulphur etc.

Layer. 1. Mature female fowl which has been kept for egg-laying purpose, especially one in current egg production. 2. Plant twig or shoot which is tied down and partially covered so that it can take root while remaining unseparated from its parent.

Layerage. Method of plant propagation by asexual means in which a portion of stem, shoot or branch has been covered with soil or some other medium in which root can develop, after which the rooted portion gets detached from the parent plant.

Layering. 1. Part of the process of hedging in which partly cut stems are being over and woven with others to produce a new hedge. Also called plashing or pleaching. 2. The pegging in the ground of a branch or half-cut branch from the stem of a plant so that it produces new roots and gives rise to a new plant. A common method of propagation.

Laying Period. The period of season during which polutry are in lay normally of about 50 weeks duration, and usually commencing at 22-24 weeks of ege for egg-producing strains, and 26-28 weeks for Broiller strains. Most egg-producing fowl are slaughtered after the first laying period as fewer eggs are laid in the second period.

Lea. Open country such as meadow, pasture, or arable land left Fallow or under grass. Another term for ley.

Lea Plough. A type of plough with a long pointed share, keeping the cutting edge well in front of the slightly twisted convex mouldboard, thus producing a continuous smooth furrow slice, as distinct from the rough broken slice of the Digger Plough. This type only ploughs to about 15 cm (6 in.) depth and is rarely used now.

Leaching. The removal of nutrients in solution from the soil, particularly under heavy rain.

Leaf Roll. A virus disease of potatoes transmitted by Aphids causing the leaves to curl upwards and inwards, to thicken and turn yellow, and the plants to remain stunted. Yields are reduced by up to 90%.

Leam. An open drain, particularly used in Land Drainage systems in the Fens.

Legbar. An Auto-Sex Linked breed of poultry produced by crossing the brown Leghorn and the borred Rock. Barred in colour with yellow legs, a single comb, and producing white eggs.

Leghorn. A small, nervous, laying breed of poultry, variously coloured including black, blue, brown, black and white, gold and silver, and white, the last being the most popular. Yellow-legged, single-combed, and producing large numbers of white egg.

Legume. A plant of the pea family (Leguminosae) which produces seed in pods and is characterised by five-pettalled flowers and by root nodules capable of Nitrogen Fixation. Some are cultivated for their protein-rich seeds (*e.g.*, Peas and Beans) for human consumption and for feeding to stock. Others are sown with grasses to produce mixed leys (*e.g.*, Clovers, Lucerne, etc.) Legumes are often grown as a break between cereal crops to enrich the nitrogen in the soil.

Lehmann System. A system of pig feeding designed to utilise relatively cheap roots or other bulk foods in quantity. The pigs are fed a basic fixed ration of Meal (a mixture of barley and fish meals) and, as they grow, increasing amounts of roots, mainly cooked potatoes, according to appetite.

Leicester. Liecester Longwool. English Leicester.

Lemma. The lower of the two bracts which enclose the flower of a grass.

Leptospirosis. A bacterial disease (*Leptospira hardjo*) of cattle which causes abortion, results in serious masttis and a dramatic loss of milk. It is transmitted via the urine and can also affect man. It is estimated that 30% of cattle herds in southern Britain are affected, and up to 90% in bad areas like South Wales. Another strain of the bacteria, *Leptospira australis,* can cause infertility in pigs.

Let Down. The release of milk from a cow's Udder during milking activated by the hormone oxytocin, the secretion of which can be stimulated by various events such as washing the udder before milking, introducing feed to the managers, etc. It is a onditioned reflex. Let down normally lasts about 7-10 minutes after which milk extraction from the udder becomes increasingly difficult.

Level. In general terms a flat, usually low-lying area of land. Specifically applied to a Drainage District or lowland area in a river valley Fen with more or less uniform drainage characteristics and usually managed as one unit, *e.g.*, Pevensey Levels, Romney Marsh.

Lewing. A screen of material, *e.g.*, coir mesh, plastics-coated netting or plastic strips, erected on the exposed boundaries of hop gardens and Orchards where natural windbreaks do not exist.

Ley. Land temporarily sown to grass (Grassland Species) and ploughed after 1-3 years (short duration ley) or after a longer period, up to 10 years in some cases (long duration ley). Seed mixtures are available to suit particular conditions and the type of ley required. Leys may be established by (*a*) Undersowing, either by broadcasting (Broadcast) or preferably by drilling (Drill), (*b*) Direct Seeding, (*c*) Direct Reseeding, sometimes with a Companion Crop, to encourage early grazing, and (*d*) direct drilling into old grassland previously treated with a herbicide. Short duration leys are an important element in crop Rotation.

Ley Corn. A cereal crop which follows a Ley in a Rotation.

Ley Farming. A system of farming in which several fields (and sometimes an entire farm) are cropped as Leys with regular reseeding, or, alternatively any rotation or cropping system which includes leys. Also called Alternate Husbandry.

Lichen. A lowly type of plant comprising an alga and a fungus living together symbiotically. Commonly found on tree trunks, old walls, bare rock, etc. Lichens are primary colonisers of rocks and important in soil formation.

Lift. To harvest root crops and potatoes, etc., by digging them from the ground, either mechanically (Potato Harvester, Sugar Beet Harvester) or by hand, ass distinct from harvesting by cutting.

Lifting Shares. A device commonly used on a Sugar Beet Harvester, consisting of a pair of triangular-shaped steel Shares, which lift the beet from the ground onto the elevator section to be deposited in a trailer.

Light. Livestock are said 'to go light' in various ways, *e.g.*, (*a*) a dairy cow when a teat dries up, (*b*) an animal when it becomes lame in one foot, (*c*) a fowl when it develops tuberculosis.

Light Soil. Soil with a high proportion of sand, a low drawbar pull, and easier to cultivate than a Heavy Soil, and thus sometimes called boy's land.

Light Sussex. A variety of Sussex poultry characterised by white feathers and black marks on the neck, wing tips and tail. A dual-purpose breed producing good quality white meat and brown eggs.

Lignin. A complex organic substance deposited in plant tissue, particularly in woody plants, and often combined with cellulose, giving rigidity and strength to stems and tree trunks. It comprises about 25-30% of the wood of trees.

Lime. Various Calcium containing materials applied to the soil to raise pH and correct acidity (Acid Soil) and mainly derived from natural deposits of Chalk and Limestone. The main types include Ground Limestone, ground chalk, Burnt Lime, hydrated lime (Calcium Hydroxide) and various by-products and waste materials mainly containing Calcium Carbonate. Applying lime to soil improves soil structure, provides calcium (and sometimes magnesium) for plant nutrition, and makes other nutrients (*e.g.*, Nitrates, Phosphates and Potash) freely available to plant, although if over-supplied same minor nutrients (*e.g.*, Boron, Manganese) can be made unavailable.

Lime Requirement. The amount of Lime needed to raise the pH in the top 15 cm (6 in.) of soil to about 6.5. It is expressed in tonnes per hectare (or cwt per acre) of calcium oxide or calcium carbonate. The amount varies according to the acidity and type of soil, *e.g.*, Heavy Soil require more lime than Light Soils. (Neutralising Value).

Limestone. 1. Various types of sedimentary rock with a high calcium carbonate content, as much as 99% in the case of chalk. Most types have an organic origin, particularly chalk, containing the hard calcareous remains of small marine organisms. Some types are relatively hard, *e.g.*, the carboniferous limestone of the Derbyshire Pennines; other are softer, *e.g.*, the magnesian limestone of Yorkshire. All are soluble in water containing carbon dioxide. Limestone is an important source of various types of lime. 2. A moorland breed of sheep, indigenous in the Southern Pennines, which contributed to the development of the Derbyshire Gritstone. Now numerically unimportant.

Limousin. A French breed of beef cattle recently imported into Britain. Red in colour with somewhat lighter tanned legs. Large bodied, rapid growing, and noted for producing a carcase with a high meat to bone ratio.

Linch Pin. One of the pins used to lock an implement onto the three-point-linkage of a tractor.

Lincoln Curly Coat. A now extinct breed of pig with a curly white coat, lopped ears and a heavy face which contributed to the development of the Chester White.

Lincoln Longwool. The largest and heaviest breed of sheep in Britain. Noted for its long shiny wool which grows over the forehead and falls over the eyes and is much used for lustre yarns and sheepskin rugs. A polled, white-faced breed, which has been widely exported.

Lincoln Red. A large breed of uniformly deep-red coloured cattle developed in Lincolnshire, and closely related to the shorthorn, with short, forward and downward curling horns and a short, broad head and with a long muscular body and short stocky legs. Formerly regarded as a dual-purpose breed, it is now popularly crossed with dairy breeds to produce beef calves.

Line Breeding. The mating of related animals, but not tnose closely related such as cousins or parents, with off-spring.

Liner. A synthetic rubber tube which lines the teat cup of a milking machine.

Linkage. The association of two or more Genes, usually near each other on the same chromosome, so that they tend to be inherited together. The characteristics passed on by such genes are said to be linked.

Linnaen System. A binomial system of naming plants and animals developed by the Swedish naturalist, Carl Linnaeus, in which the first name represents the Genus and the second the Species, *e.g., Hordeum sativum* (barley).

Linseed. Varieties of the plant *Linum usitatissimum,* selected and grown for their oil-rich seeds, which are sometimes fed after crushing, or after scalding or cooking, particularly to young calves and sick animals, but more usually as linseed cake which puts a 'bloom' on fattening cattle.

Lipase. An Enzyme which splits fatty acids into alcohol and acid. It is present in Intestinal Juice and Pancreatic Juice.

Liquid Manure. The urine of farm animals, Dung Liquor and drainings from farm buildings containing valuable Nitrogen and Potash, at one time mostly collected in tanks and distributed on fields or over solid dung heaps. Nowadays urine and faeces are commonly mixed with litter to produce Farmyard Manure, slurry is another type of liquid manure.

Liquid Nitrogen Fertilizers. Various nitrogen fertilizers applied in liquid form, usually less concentrated than the granulated, powder and crystal forms, *e.g.*, a solution of urea and ammonium nitrate containing 25-30% nitrogen, and often applied as a top dressing, particularly to grassland and winter cereals. Other nitrogenous fertilizers used in liquid form include gas liquor and pressurised anhydrous ammonia.

Litter. 1. Bedding for livestock, *e.g.*, straw, shavings, sawdust, etc. 2. All the young animals born to a female at one time, particularly applied to pigs. The average litter size is 11 piglets, but can be as many as 20 or more, and over 10% die in their first few days due to cold or crushing. Wel managed breeding sows produce between 2.0 and about 2.6 litters per year depending on the age of weaning of piglets.

Liver Fluke, Liver Rot. A parasitic flatworm (*Fasciola hepatica*), the common fluke, which infests the liver of various animals, particularly sheep and cattle, causing bile duct inflammation, diarrhoea, wasting, and reduced milk yield in dairy cows. Eggs pass out with the faeces, hatching to produce larvae which infect various species of mud-snails (the alternative Host), and after multiplication they leave the snails to form cysts on grass, to infect grazing animals. A very widespread disease of cattle and sheep. Also called bane and cord.

Live Stock. Domesticated animals such as cattle, horses, pigs, poultry and sheep, etc. Defined in the Agriculture Act 1947 as including 'any creature kept for the production of food, wool, skins or fur, or for the purpose of its use in the farming of the land'.

Liveweight. The weight of a live animal as distinct from the weight of the Dressed Carcase of the animal.

Loam. A soil with a balanced soil particle mixture (approximately 25% Clay, 40% Sand and 35% Silt), having most of the advantages of Clay Soils and Sandy Soils and few of their disadvantages. Loams are easily cultivated, but can be sticky when wet, and are usually drained.

Local Lesion. Localized leaf spot appearing as a result of mechanical inoculation with a virus.

Locks. A type of wool oddment purchased by the British Wool Marketing Board comprising small pieces of wool detached from a fleece during shearing.

Load. An open Ditch or Drain in the Fens.

Lodged. Laid Corp.

Long Dung. Farmyard Manure.

Longhorn. A hardy breed of beef cattle, characterised by its long, forward and downward curving horns. Coat colour is variable including roan, red and dark plum, and many intermediate shades. There is usually a white ridge along the spine and tail, and white is also normally found on the face and thighs. A long-bodied, short-legged breed, producing a lean carcase, and milk high in buttterfat, at one time much used for cheese making.

Long Manure. Another term for long dung.

Loose-box. A stable or part of a stable in which animals are kept united with no other fittings other than a hay-rack, manager, water-bowl, and tying ring(s). Often several are grouped together in a row with two-halved doors, and insulated concrete floors with a slight drainage slope. Conveniently used for isolating diseased animals

Lop Ears. Ears which hang loosely and often cover the eyes.

Louping III. A paralytic virus disease of sheep carried by a Tick (*Ixodes ricinus*), common on mountain pastures, particularly in northern England and Scotland. The sheep are caused to twitch, quiver, and stagger, and eventually to leap erratically before paralysis occurs. Also called staggers.

Louse. A wingless, flat bodied, parasitic insect, with short legs, found on the skins of livestock, mostly sucking blood, *e.g.*, Ked (sheep louse).

Lowland. Low-lying land in a region or district as distinct from hill areas and mountains.

Low Loader. A trailer with its bed near to the ground facilitating easy loading.

Lucerne. A perennial leguminous plant (*Medicago sativa*) grown for both Fodder and Forage. Characterised by deep roots (with Nodules) giving it drought rêsistance, trifoliate leaves, distinct purple flowers, and twisted pods containing kidney-shaped seeds. It is intolerant of acid or poorly drained soils. Also called alfalfa.

Luing. An early-maturing, hardy breed of beef cattle, recently developed from crosses between the Beef Shorthorn and Highland breeds. Red coloured with some roan, gold or white.

Lupins. A leguminous plant (*Lupin sp.*) sometimes grown as a green manure crop for ploughing in, particularly on poor light land in the eastern countries deficient in lime and humus. White, yellow and blue flowered varieties exist. The sweet lupin has been used for forage, being the only non-toxic variety.

Lymph. A watery colourless liquid derived from blood filtering through tissues from capillaries and collected in lymph vessels to be returned to the bloodstream. It carries nutrients to the body's tissues.

Lynchet. A boundary ridge or unploughed strip. Also a terrace or bank formed on a slope due to continuous contour ploughing. Also called linch.

Lysimeter. A device which is used to measure the quantity or rate of downward water movement through a soil usually undisturbed, or to collect such percolated water for analysis as to quality.

Lysolfecithin. One of the substances in the venom of bees and wasps: a very toxic substance which breaks down the cells of its victim an sets free histamine.

Lysozyme. Ezsyme that catalyses the destruction of the cell walls of many bacteria by hydrolysing the mucopeptide.

Lytic. Pertains to the dissolving of a substance *e.g.*, the dissolving of certain bacteria by lysozyme.

M

M_1. Generation which follows mutation or mutagenic treatment.

Macro-Elements. Macro-nutrients. Elements which are required and found in relatively large quantities as natural constituents of organisms or tissues; major elements, like nitrogen, phosphorus and potassium.

Macroenvironment. The term used for the environment due to general, regional climate, traditionally measured some 4 ft above the ground and away from large obstructions.

Macronutrient. A chemical element needed by crops in relatively large amounts (normally geater than 1 ppm), normally applied as a fertilizer or liming material.

Maedi-Visna. A slowly progressive virus disease of sheep and goats, first reported and named in Iceland (maedi' laboured breathing; visna' wasting) in 1939, and now widespread symptoms include slight distress. Infected sheep will lag behind others, breathing is fact and shallow, becoming progressively more laboured until the animal fails and dies after 3-8 months. No vaccine or treatment is available. Lambs are infected from ewes via milk, and the disease also spreads readily by close contact between animals.

Maggot. A legless larva of certain two-winged flies of the order Diptera, with a very tiny head. The maggots of many flies attack livestock and crops, e.g., Blow Fly, Crane Fly, Warble Fly, etc.

Magnesium. (Mg). A metallic element found in many rock types (e.g., dolomite, magnesite) and essential in life for metabolism. A constituent of chlorophyll. Deficiency in the soil can cause Chlorosis in plants and can lead to gass staggers in livestock, and is corrected by applying magnesian limestone (Ground Limestone), Epsom Salts or Kieserite, etc. Other Fertilizers containing magnesium include farmyard manure, kainit and basic slag.

Maid. A ewe teg shown to be barren after being served by a ram.

Maiden Seeds. A ley which has not been grazed or mown.

Maiden Tree. A fruit tree, about 1 year old. Also a fruit tree developed from a seed as distinct from one developed from a graft.

Main Canal. In irrigation practice, it refers to a canal which extends from the source of water supply to the beginning of the distrubution system.

Maintenance Ration. The amount of food which when supplied to an animal will keep it healthy and maintain a constant live weight, with no loss or gain.

Maintenance Requirements. Portion of the energy of the food which is used in the life process of the animal for keeping the animal warm and for movement of the body.

Malaria. Febrile disease of animals which is caused by the presence of certain protozoan parasites in the blood for gain.

Maize. A cereal plant (Zea Mays) now commonly grain, silage, forage and fodder. It requires good drainage, grows best on lightish soils and is frost-sensitive. Characterised by a thick, stout, tall stem, and numerous long leaves arising alternately on each side of the stem. The male flower (tassel) forms at the top of the shoot, and several female flowers form in the axils of some of the middle-stem leaves and develop into long cylinders (cobs) of tightly packed gain. Maize is rich in Carbohydrate but low in utilisable Portein and Fibre. It is much used as a Meal for fattening livestock but is also fed flaked. Also called corn on the cob.

Malathion. An organophosphorus insecticide and acaricide available either alone or in mixtures (i.e., with D.D.T.). Used in liquid or dust from, particularly to control skin parasites of cattle, and to control various crop pests.

Malignant. Cells which are dividing and enlarging autonomously.

Malling-Merton Series. Group of apple root stocks which were the result of crossing.

Malt Culms, Coombs. A by-product of the Malting process, used as a Feedingstuff for dairy cows, sheep and horses, after thorough soaking.

Malt Sugar. Maltose.

Maltase. An enzyme which breaks down maltose into glucose. It occurs in yeast and is an important element of the brewing process.

Malting. The process whereby barely or other gain in soaked in water and allowed to partly germinate (during which stored Starch is broken down to Maltose by the enzyme Diastase), and is then kiln dried, after which the roots and shoots are removed (malt culms) and the remaining grain (the malt) is used in brewing. The malt is steeped in water and the maltose extracted into solution (the wort) which is run-off, leaving the residual grain which is used as an animal Feedingstuff.

Malting Barley. The best quality Barely sold for malting, representing about one fifth of the crop, for which a premium price is paid. It has a high Starch content (required for obtaining Maltose) and a low Protein content (determined during testing by measuring Nitrogen content). Good quality malting barley, after being harvested well, should be dry, and should smell 'sweet' (with no suggestion of mustiness). Also the grain should be evenly size with few broken or damaged ones, evently coloured wite wrinkly surface, and free from weed seeds and disease. These qualities are important for malting, and tests with strict criteria are carried out by millers before grain is accepted from fermers.

Maltose. A disaccharide sugar (Carbohydrate) comprising two Glucose molecules, formed by Starch breakdown by the enzyme diastase, particularly during and seed germination (Malting). Also called malt sugar.

Mamma. Milk secreting organ of female mammals.

Mammals. A class of animals characterised mainly by hair, and the secretion of milk to feed the young (See Mammary Glands), which includes all farm livestock, except poultry.

Mammary Glands. Milk-producing glands, characteristic of female Mammals. Located in the Udder of cows. Milk production and secretion is controlled by various Hormones. (See Lactation).

Mane. 1. Long hair on the back of a horse's neck. 2. A narrow strip of grass or cereal not cut by a mower, or by a Binder or Combine Harvester.

Manettii. Name given to a strong vigorous type of rose which is often used as a stock in the propagation of cultivated rose varieties.

Manganese. (Mn). A metallic Trace Element required for metabolic processes as an enzyme activator, it can be deficient in Neutral or Alkaline Soils rich in Organic Matter, causing Deficiency

Diseases in crops (e.g., Speckled yellows in beat, Grey Leaf in cereals), which is usually rectified by spraying with manganese-sulphate. Cattle infertility may result if it is deficient in pasture.

Mange. A skin disease of animals caused by mites which results in the hair dropping out in patches or, due to persistent itching, being rubbed away by the animal.

Manger. A trough in which food for cattle and horses is placed, often located in a Cowhouse or Stable.

Manioc. A tropical plant (*Manihot esculenta*) producing a tuber with a high Starch but low Protein content, used as a cereal substitute in animal feedingstuffs. Also called cassava or tapioca.

Manure. A general term for anything added to the soil to increase its content of plant nutrients (Fertilizers). Normally applied to organic, usually bulky, materials mostly derived from farm and animal waste products, such as Compost, Dung, Farmyard Manure, Slurry, Sewerage Sludge, etc. They are slower acting than inorganic fertilizers but have the advantage of adding Organic Matter to the soil, increasing its capacity to retain water, and improving its structure (Green Manure).

Manure Spreaders. Various types of machine used to distribute solid or liquid Manures on the land. Also called muck spreaders. The main types include:

(*a*) Self-emptying trailers containing a slatted conveyor in the base, delivering the (mainly solid) manure to a revolving mechanism at the rear which shreds and flings it on the field. Also called trailer-spreaders.

(*b*) Cylindrical, open-topped tanks with a p.t.o. driven, central, axial shaft which operates a set of attached flail chains, which throw the manure (usually containing Slurry) out to one side of the tank.

(*c*) Gun mechanisms which distribute, under pressure, slurry conveyed via a pipe from a nearby fixed tank or from a mobile vacuum tanker. The latter are sometimes used to spread slurry directly onto the land.

Manurial Residues. That proportion of a manure added to the soil but not used by the crop, or residues left in the soil by a crop, which are available to future crops.

Manurial Value. Those nutrients in a feedingstuff which an animal does not take in and pass out with the faeces so as to enrich the dung. Also called the residual value or fertilizer value of a feedingstuff.

Maran. A heavy breed of poultry with barred feathers, white legs, a single comb, producing dark brown eggs. Also used as table birds.

Marden Layer. A term for a young Ley.

Mare. A female horse of 5 years of age or more.

Marek's Disease. A virus disease of poultry causing lameness, progressing to leg paralysis drooped wings, neck twisting and greying of the iris in both eyes.

Marginal Land. Poor quality land suffering from various permanent disadvantages such as steep slopes, bad drainage, poor thin soil, harsh environmental conditions, etc.

Marked Coat. An animal's coat having two or more specific colours as distinct from a single colour (self-coloured).

Market Garden. Land used to grow produce, mainly vegetable and fruit, for sale (usually to a market, e.g., Covent Garden Market) for direct human consumption, as distinct from for processing purposes. The term is often restricted to the production of vegetables on a small scale.

Marl. Calcareous clay which is used as a dressing to improve soils particularly through the addition of clay to light soils.

Marmalades. These are made usually from citrus fruits and consist of jellies of the concerned fruit containing shreds of peels suspended in them. A good jelly marmalades can be prepared by using sweet orange (Malta) and Khatta (sour orange) in the ratio of 2 : 1 by weight, or Santra (locse orange) and Khatta 2 : 1 by weight.

Marsh. An area of usually low-lying land, often flooded by fresh or saline water, with poorly drained, impermeable mineral soil usually silt or clay (as distinct from the organic soils of Fens) which is continuously waterlogged. Characterised by sedges and rushes. Many marsh area have been artificially drained and protective embankments established to prevent flooding, with subsequent improvement of their grazing value, and in some cases the introduction of arable cropping.

Mash. A mixture of animal foods, crushed and stirred to a soft pulpy state with water (wet mash) or without it (dry mash).

Mashlum, Maslin. A mixture of oats and barley and sometimes wheat, sown to provide grain for feeding to livestock. Occasionally beans or peas may be added to the mixture but ripening in uneven and they are not suited to being harvested by a Combine Harvester. Also called dredge or meslen.

Mast. The fruit of Beech, Chestnut, Oak and other trees, eaten by pigs as Pannage.

Mastitis. Inflammation of a cow's Udder, due to bacterial infection, resulting in the secretion of clots of milk, and sometimes pus when the infection is acute. Also called mammitis, garget and udder-clap

Mat. A closely interwoven or tangled mass of dead grass which develops when a pasture is undergrazed, particularly on acid soils.

Maw. The stomach, particularly the fourth stomach of Ruminants.

Mayflies. Common name for the Ephemeroptera.

Mazzard. Wild cherry which is used extensively as a rootstock upon which commercial varieties of cherries are grafted.

M.C.P.A. An abbreviation for 2-methyl-4-chloro-phenoxy-acetic acid, a Translocated Herbicide or Hormone Weedkiller, used to control many broad-leaved annual and perennial weeds. Often also include in mixtures with other translocated herbicides.

M.C.P.P. An abbreviation for 2-methyl-4-chloro-phenoxypropionic acid. A hormone herbicide used mainly to control weeds in cereals. Also called mecoprop.

Mead. 1. A shorter version of meadow. 2. Wine produced from honey.

Meadow. Strictly applied to an area of grassland used to produce Hay or Silage rather than for grazing. Also used loosely for rich waterside fields which are frequently flooded by river water, either naturally or via sluices and which are grazed (water meadows).

Meadow Fescue. A loosely tufted perennial grass (*Festuca pratensis*) forming large russocks, with narrow, bright green hairless leaves, shiny on the underside, and loose, green or purple panicles. Valuable for hay and grazing, and particularlyfound on rich, well-drained, fertile clay soils, and often abundant in water meadows. A very palatable grass which grows well mixed with white clover or timothy.

Meal. 1. A type of feedingstuff consisting of a single or a mixture of finely ground ingredients, e.g., cereals, oil seeds, meat, bone, fish, etc. Mainly used for pigs and poultry. 2. Another term for oil seed cake.

Measles. A condition of the carcases of cattle and pigs characterised by presence of Tapeworm cysts in the meat.

Measure-Weight Test. A test which gives an indication of the relative heaviness, pulpness and quality of seeds. For this purpose, a specified volume of seed is weighed and represented as kilos per hecto litre.

Measuring Weir. Device or structure which is used for measuring the flow of water. It generally consists of a broad-crested weir or a rectangular, trapezoidal, triangular, or other shaped notch, frequently located in a vertical, thin plate over which the water flows. The weir head is an index of the rate of flow.

Meat. Flesh of animals which is used as food. It consists of proteins, fats, minerals, vitamins, and water. The most common meats used as food are lamb, beef, and pork. Poultry and fish are sometimes included with the foods classified as meat.

Meat Meal, Meat-and-bone Meal, Meat Guano. Meal derived from waste meat and bone trimmed off animal carcases and steam cooked until moisture content is below 10%. Surplus fat is drained off and most of the rest expelled by pressing the residue. By law meat meal must contain at least 55% protein, meat-and-bone meal 40-55% protein. In both types salt content must not exceed 4%. Also called tankage.

Meadian Lethal Does (MLD). Does of radiation which is required, to kill, within a specified period, 55 per cent of the individuals in a large group of animals or organisms.

Median Lethal Time (MLT). Refers to the time required, following administration of a specified dose of radiation, for the death of 50 per cent of the individuals in a large group of animals or organisms.

Melanin. Insoluble black pigment which is laid down in the cuticle of many insects as well as in the skins of a great many other animals; formed by oxidation of the amino acid tyrosine in the presence of specific melanogenic enzymes.

Melanized Soil. A soil having a very dark colour to a considerable depth in the surface layer, due to the incorporation of humus.

Melitten. Poisonous protein like substance which is present in the venom of bees.

Mercerize. Refers to the process for treating cotton fibres, yarns, or fabrics with a solution of concentrated caustic alkali to make them stronger. If kept under tension during the process they acquire a luster.

Merino. Breed of small sheep which is producing wool of excellent quality, very fine and of 4 inch staple, 'the golden-flooted sheep', with its ancestral home in Spain; now reared in large numbers in Australia, South Africa and USA. Yields finest fleece, clip varying 6 to 30 *lb* a year.

Metabolite. Chemical participating in metabolism e.g.a nutrient.

Metered Chop Harvesters. Two kinds of forage harvester, producing a uniform short chop, having either a flywheel or cylinder type of chopping mechanism. With both types the chop length and rate of delivery by a feeder to the chopping mechanism can be controlled. These harvesters are usually trailed, but self-propelled machines are available. Also called precision chop harvester.

Methane. (CH_4). A colourless hydrocarbon gas which can be formed in large quantities in a cow's Rumen. Now being developed as a source of fuel on farms, produced by fermenting slurry.

Meuse-Rhine-Ijssel (M.R.I.). A dual-purpose breed of cattle, taking its name from three rivers in Holland where it was developed. The breed has a similar milk yield to the Friesian but better meat production. A large-bodied breed, characterised by a patched red and white coat, with short white legs, white belly and tail, and sometimes a white blaze on the broad head, and by short, forward and inward curving horns.

Micro Budding. Method of budding which is commonly used in propagating citrus trees. A bud is removed from the bud-stick by giving a flat cut just underneath the bud with a sharp knife after cutting off the leaf petiole just above the bud. An inverted T-cut is made in the stock wherein the micro-bud is slipped keeping the right side up. The entire inverted T-cut including the bud is covered and exposed only after 2-3 weeks so that healing gets completed.

Microbe. Microscopic organism belonging to either the animal or the plant kingdom.

Microbial Insecticide. Pathogen which is used for killing insects. Free from harmful residue, nonphytotoxic to crops of higher dose, harmless to beneficial insects.

Microbiology. Science and study of microscopic organisms.

Microenvironment. Environment which is close enough to the surface of a living or non-living object to be influenced by it.

Microingredients. Added vitamins, trace minerals, antibiotics, and other materials used in minute amounts.

Microplot Thresher. Self-cleaning machine thresher operated by a gas engine or by electricity for threshing the grain from microplots in rapid succession without mixing the seed of two varieties.

Midden. A Dung heap.

Mids. Potatoes sold for human consumption, which are of 'Ware' standard, and which will not pass through a 20 mm horizontal mesh when manipulated, but will pass through a 45 mm mesh.

Milch Cow. A cow giving milk or kept for milking.

Mildew. Various fungi, which are parasitic on plants, causing disease. The threads or hyphae of some Phycomycete fungi penetrate deeply (Downy Mildews), whilst other types of Ascomycete fungi grow more on the leaf surface with a powdery white appearance (Powdery Mildews). Also called moulds.

Milk. A white liquid secreted by female mammals (Mammary Glands) for the nourishment of the young, containing valuable essential food constituents, e.g., Proteins, Carbohydrates, Fats, etc. (Milk Composition). Milk produced by dairy cattle for liquid consumption and for the manufacturing of various products (e.g., butter, cheese, yogurt, etc.) is marketed under the control of the Milk Marketing Board. Whole, fresh milk sold for public consumption must contain at least 3.0% butterfat and is tested for quality on purchase from the farm quality scheme. Several types of liquid milk are available including Untreated Milk. Pasteurised Milk (including Homogenised Milk), Sterilised Milk, and Ultra Heat-Treated Milk.

Milk Clipping. The clipping of ewes which are still suckling lambs.

Milk Composition. Cow's milk is a balanced food containing all the nutrients (including vitamins) necessary to humans and is particularly fed to children and young animals. The composition can vary according to breed (Cattle), age level and type of feeding, stage of Lactation, etc. The following are average figures :

	Per cent	
Protein	3.40	..........mainly Casein
Sugar	4.75	..........mainly Lactose
Fat	3.75	..........Butterfat
Mineral Matter	.75	..........Ash
Water	87.35	
	100.00	

Milk Feyer. A disease of various animals, particularly dairy cows, usually occurring a few days afters calving. It is associated with a fall in the blood calcium content the reasons for which are not fully understood. It results in initial excitement, then loss of balance and later of consciousness, with the neck in a typical twisted position. Not an actual fever, body temperature often falling a few degrees below normal. Treated by injecting calcium borogluconate.

Milk Recording. The recording of the milk yield of each cow in a dairy herd by measuring the volume of weight of milk given during milking sessions. Usually the quantity can be read off a scale on the collecting jar into which the Milking Machine initially extracts the milk.

Milk Sinus. The cavity in each of the teats of a cow's Udder into which milk is secreted and from which it is extracted during milking.

Milk Vein. One of two large veins visible on a cow's belly running from the udder.

Milk Vetch. Various leguminous plants (*Astragalus spp.*) grown on lime-rich soils for fodder and supposed to increase the milk yield of dairy cattle.

Milking. The extraction of milk from a cow's udder by hand or milking machine Milk secretion is actuated by the let down mechanism, and the squeezing action of the hand or the squeezing-suction action of the teat cups withdraws the milk from each milk sinus. (Milking Parlours).

Milking Machines. There are three main types of milking machines all of which, through a pulsator mechanism, apply an intermittent partial vacuum to the teats of the cow's Udder, simulating the sucking action of a Calf. They vary in the design of collecting the extracted milk and include (*a*) bucket plants, used mainly in cowshed milking, in which milk is extracted into covered buckets and carried to the milk cooler, (*b*) in-churn plants, in which the milk is conveyed directly into churns in which it is cooled and transported to the dairy (see Churn Collection), and (*c*) releaser plants, in which the milk is extracted into a collecting jar (in which each cow's yield is quantified) and from which it is sucked via vacuum piping to be 'released' into the cooler, but without disturbing the vacuum system. This system is now almost universal. The abbreviation m/m is sometimes used for milking machine.

Milking Parlour. A building containing specialised milking machines. Many designs exist, the choice depending on herd size, average yield, available labour, feeding method, etc. There are two main categories of parlour, having either one or two stalls per milking unit, the latter having a greater through-put at about 10 cows per unit per hour. Concentrates are often automatically fed to the cows whilst in the stalls. Layout may be

(a) Abreast, with stalls side by side, sometimes slightly raised.

(*b*) Tandem, with stalls end to end around a pit, so that cows stand head to tail. In the simplified chute type there are two stalls to each side of the pit and cows enter via steps or ramps and leave together after milking.

(*c*) Herringbone, in which there are no stalls and the cows stand side by side at an angle and to each side of a central pit.

(*d*) Rotary, in which any of the preceding formations move round the milker on a raised, slowly moving platform.

Mill. A machine or a building equipped with machinery for grinding (by crushing) cereal grain or other dry feedingstuffs to produce flour or meal (Crushing Mill, Hammer Mill, Plate Mill). Also to remove seed husks.

Milling Wheat. The best quality wheat purchased from producers by the flour-milling industry for bread making, etc. Such wheat, normally grown under contract, must reach specified standards, and samples are subject to visual and qualitative test before acceptance by the miller, *e.g.* Grain Moisture Content, cleanliness and purity, damaged and diseased grain content, odour, bushel weight, protein content, Gluten quality, Aphaamylase content, etc.

Million Reaction. A test for presence of tyrosine in the molecule. The Million reagent consists of mecurous nitrate in nitrous acid. Four to five drop of this reagent are added, mixed and gradually heated with 5 ml of solution. A white precipitate is formed which turns red on standing. The reaction fails to occur if the protein is not precipitable by strong acid.

Mineral. A term mainly applied to inorganic substances, usually occurring naturally in the soil or rocks as distinct from those substances which are organic, being derived from living matter. (Mineral Matter.)

Mineral Matter. Mineral elements and compounds, found as a constituent of the soil and living organisms, and essential to metabolism. Mineral matter in the soil consists of fragments of rock in various stages of degradation and their decomposition products, e.g., quartz, clays, iron oxides, etc., intimately mixed with organic matter. Some of the mineral elements may be present in plants and animals as mineral salts or combined in organic compounds such as proteins. Certain parts of plants have a higher mineral content (e.g. roots, vegetative parts), and as green crops mature the mineral content of the Dry Matter diminishes. The useful mineral matter content of feedingstuffs (expressed as Ash) of animal origin is normally higher than in those of vegetable origin.

Mineral Mixtures. Minerals including Trace Elements, supplied to animals to ensure an adequate supply for metabolism and to prevent deficiency diseases, infertility, lack of thrift, or poor production etc. Often added to concentrate mixtures or cakes. Also given loose as mineral mixtures in field troughs, or as hard slabs (mineral licks or salt licks) for livestock to lick at will.

Mineral Phosphate. *See* Rock Phosphate.

Mites. A sub-class of the Arachnida allied to ticks, found in many habitats (e.g. the soil decaying Organic Matter) and parasitic on both plants and animals.

Mixed Cropping. System of sowing two or three crops together on the same land, one being the main crop and the other one or two subsidiary. Has some advantages as of following crop rotation: less of risk of total crop failures, better soil utilization and satisfying the requirements of the farmer in food and fodder. Common mixtures followed in India are upland rice, cotton and red gram, jowar with *Sateria italica, Panicum miliare and Paspalum scrobiculatum,* bajra with fibres like mesta and sannhemp, ragi with ginger, red gram, cowpea and jowar.

Mixed Farming. Type of farming under which crop production is combined with live-stock raising. The live-stock enterprise, is complementary to crop production programme so as to provide, a balanced and productive system of farming. A farm to be categorised as of mixed type, at least 10 percent of its gross income must be contributed by the live-stock activities, the upper limit being 59 per cent.

Mixed Fertilisers. Consists of individual or straight fertilizer materials blended together to permit application in the field in one operation. Mixed fertilisers supply two or three major plant nutrients. The percentages of these nutrients are expressed as fertiliser grade.

Mixen. A Dung heap. Also called a midden.

Mixed Grazing. Two or more classes of livestock, such as sheep and cattle grazed on the same pasture.

Moisture Penetration. Depth to which water penetrates in a soil following irrigation or rain; the downward movement of water in a soil.

Moisture Percentage. Mode of expressing moisture in the soil; the ratio of the weight of water in the soil to the dryweight of the soil, and expressed as a percentage of dry weight.

Moisture Release Curve. Depicts the relationship between soil moisture tension and the absolute moisture content in the range from field capacity to wilting point.

Molasses. A thick treacle produced as a by-product of sugar refining, containing 20-30% water, and rich in soluble sugars (particularly

Sucrose). Used to improve the palatability of Feedingstuffs, and as a binding agent in the production of Cake. Also used as a laxative and as a wilting agent in Silage making.

Mole Plough. An implement used to form mole drains, consisting of a torpedo-shaped cylinder (a mole or catridge) attached to a vertical, knife-edged coulter, which is itself fixed to a metal beam suspended between two wheels. Attached directly behind the mole is an expander (a tapered cylinder of greater diameter) which slightly enlarges the drain being cut, smoothing and compressing the drain walls as the mole is drawn through the soil.

Molisch Test. A biochemical test for carbohydrate. A few drops of α-naphthol are added to a solution or suspension of the suspected carbohydrate and concentrated sulphuric acid is slowly poured down the side of the test tube. A violet ring at the juncture of the two solutions indicates carbohydrates.

Molluscicide. A chemical used to kill slugs and snails.

Molybdenum. (Mo). A metallic trace element required by Legumes for Nodule formation. When present in excess (e.g., the Teart soils of the West Country) it can cause scours in Ruminants, particularly cattle.

Moniliasis. A fungal disease affecting mainly the poultry. The infection is usually confined to the alimentary tract and rarely causes a generalized disease. Fungus of the genus *Candida* is the casual organism.

Monoculture. The growing of the same crop on a field year after year.

Monoecious. A term applied to plants having both separate male and female flowers on the same individual plant, e.g., hazel.

Moor, Moorland. Open country, usually on high ground, having acid soil (mostly Peat) and covered with heather, coarse grasses, sedges, bracken, etc.

Mor. An infertile acid soil containing undecayed plant remains, found particularly on sandy Heaths.

Mor Humus. A humus developing on acid, sandy heaths. The soil is too acid for an aboundant microflora to live, so the humus develops on the surface of the soil.

Morbidity. The percentage of a herd, flock etc., that becomes infected by a particular disease.

Mosaic Diseases. Various virus diseases of crops causing light and dark mottling on affected leaves. e.g., Virus Yellow.

Moss. 1. A northern term for a bog. 2. A class of small lower plants (Bryophyta) with simply constructed leaves, and lacking woody tissue.

Mould. 1. Loose soil with a good tilth. Also soil rich in humus.
2. A general term for a variety of minute fungi which produce a powdery or downy growth. Also called mildew. (Phycomycetes)

Mouldboard. A curved steel plate on the body of a plough which turns over the furrow slice. Also called breast or shell-board. Design varies according to the type of plough. The General Purpose Mouldboard is almost flat and is twisted along its length, producing a continuous almost unbroken furrow slice. (Digger, Semi-Digger and Lea Plough). When worn, the leading edge or shin of the mouldboard may be replaced.

Mow. 1. To cut grass or a forage crop. 2. A pile or stack of hay or corn, particularly in a barn. Also the place in a barn where such a stack is put (sometimes called mowhay).

Mower. An implement used to cut a grass crop. The main types are :

(*a*) Cutter bar mowers, incorporating a high speed reciprocating knife mechanism. They may be fully mounted or partially supported on wheels, and driven by p.t.o. or gearing from the wheels. The crop receives a single cut, without further laceration and is deposited in a neat swath.

(*b*) Rotary mowers. The flail type consists of flail knives mounted on a horizontal axle, which lacerate the grass, assisting drying (Haymaking), and are useful for cutting laid crops. The vertical spindle type consists of one to four vertical drums containing hinge mounted knives which rotates horizontally at high speed.

(*c*) Cylinder mowers, usually comprising a gang of three independent horizontal cylinders, each bearing a number of cutting blades. Normally tractor-drawn and used for cutting short grass. Rarely used on farms now-a-days.

Mower-Conditioner. A combination machine which both cuts the grass crop and conditions it to enhance drying. Various conditioning mechanisms are used, all of which bruise and abrade the waxy coating of the stems to increase epidermis premeability.

Mug. A wooly-faced sheep.

Mulch. Material such as straw, leaves, sawdust, grass cuttings, loose soil, etc., applied to the soil surface to protect the soil and plant roots from drying our and from the effects of heavy rain, freezing, etc.

Mulch Farming. System of a farming in which the organic residues are not ploughed into the ground, but are left on the surface.

Mulch Tillage. Tillage or preparation of soil in such a way that plant residues are left on the surface.

Mule. 1. A cross-bred sheep bred from a border Leicester ram and a hardy hill ewe. The original mule sheep, dating back to 1863, was derived from a Bluefaced Leicester ram mated with a Swaledale or Scottish Blackface ewe, and is known as the North Country Mule. 2. The hybrid offspring of a male ass and a mare.

Mull. A soil containing a good mixture of partly decayed Organic Matter, Humus and Mineral particles, and containing plentiful earthworms.

Multiple Suckling. AQ system of managing cattle in which dairy breed nurse cows are used to suckle several beef calves at a time. Between suckling sessions the cows are put out to graze.

Muriate of Potash. A fertilizer containing at least 60% potassium chloride, and as much as 3% sodium chloride or common salt. Available as a straight granular fertilizer but mostly incorporated in compound fertilizers.

Murray Grey. A breed of polled beef cattle developed in Australia in the Murray Grey Valley. A very gentle breed, silver to grey in colour. Cows have good milking qualities and calve easily, producing quick-growing calves.

N

Nagana. Protozoan disease of animals characterized by progressive anaemia and emaciation, intermittent fever and sometimes muscular and nervous disturbances. It is caused by four types of trypanosomes viz. Trypanosoma, congolense, Trypanosoma vivax, trypanosoma, brucei and trypanosome simial.

Natural Ecosystem. An eco-system that ultimately produces the climax community, which develops in the absence of any major disturbance and consists of a complex, self regulating food web of several trophic levels.

Neat. A cow, heifer, bull or steer.

Neck. A sheaf of corn.

Neck Collar. A broad strap which fits round the neck of a cow or horse, securing it in a stall.

Necrosis. (1) Death of organs of a plant, either as blight or death of tissue in localized areas, usually inside fruit add stems, or die-back or death of stems or branches, or blast or a separation of a part or of organ, or antheracnose or limited lesions. Characteristic symptom in certain plant disease (2) Death of a cell or group of cells in a tissue in animals due to various causes, such as invasion by certain microorganisms.

Necrotic. Death of the plant cells, especially when it results in the tissue becoming dark in colour.

Nematocide. Agent killing or inhibiting nematodes.

Nematodirus Disease. A disease of lambs caused by parasitic worms (*Nematodirus spp.*). They are transmitted by eggs in the faeces which later hatch in pasture to infect the following year's lambs, causing diarrhoea and lack of thrift, and sometimes anaemia.

Nessler's Reagent. Mercuric iodide-potassium iodide solution, used to analyze for small amounts of ammonia.

Nest Boxes. Boxes in which poultry kept under a system of deep litter are able to lay eggs. They may be of the single type designed to accommodate 5 birds or the communal type accommodating up to 100 birds. On large poultry farms the boxes are often arranged so that newly laid eggs roll on to a conveyor belt for delivery to a packing room.

Net Energy Value (N.E.). The Metabolisable Energy value of an animal food less that fraction lost in the form of heat. This value is theoretically a more accurate indication of the utilisable energy in a food than its M.E. value. N.E. values vary according to the physiological state of an animal, *e.g.,* whether it is being fattened or is on a Maintenance Ration, etc. The determination of N.E. values is expensive and time consuming and they are available for only a few foods.

Net Irrigation Requirement. Depth of irrigation water, exclusive of precipitation, stored soil moisture or ground water, that is required consumptively for crop production and required for other purposes, such as leaching.

Net Sown Area. Class of land under revenue classification. Represents actual extent of land sown to crops during a given year ; when cropped more than once in a year, its area is added to the net area sown, to arrive at the total area sown during the year.

Nettlehead. A virus disease of hops resulting in weak deformed Bines with little ability to climb or fruit.

Neutral Soil. A soil in which the surface layer is neither acid nor alkaline in reaction, *i.e.,* its pH is about 7.0 (strictly, neutrality is pH 7.0).

Neutralising Value. An indication of the ability of a liming material (Lime) to neutralise soil acidity compared to the effect of Calcium Oxide as a common factor. The n.v of a liming material is the number of kgs of calcium oxide which would have the same neutralising effect as 100 kg of the material.

Newcastle Disease. A highly infectious notifiable disease of poultry caused by paramyxo virus. Symptoms include reduced yield of often soft and misshaped eggs, lack of appetite, troubled breathing, nasal discharges and foul-smelling yellowy, watery diarrhoea. High mortality amongst young birds is common.

Nib. 1. One of the hand grips on the shaft of a Scythe. Also called dole, handpin, thole. 2. A wedge of Hops removed from a Pocket for examination, as distinct from a cube-shaped sample taken for valuation and sale purposes.

Nick. A term used of two particular animals found to produce consistently good offspring when mated together. They are said to nick well.

Nidget. 1. A horse-drawn hoe once used for cultivating the alleyways in hop gardens. 2. A sample of Hops.

Ninhydrin Reaction. Blue colour reaction used in the chromatographic identification of amino acids. A blue compound is formed when ninhydrin is heated with alpha-amino acids or with peptides or proteins which contain at least one free amino and carboxyl group.

Nipping. Act of removing of earheads from the plant stalk.

Nitrate Reduction. Reduction of nitrates to nitrites or ammonia.

Nitram. A granular of ammonium nitrate containing about 34.5% nitrogen

Nitrate. A general term for any nitrogenous fertilizer. The term is specifically applied to salts containing the nitrate ion (NO_3). The nitrogen in many compound fertilizers is in the form of the ammonium ion (NH_4), which is quickly changed to the nitrate form by soil bacteria (Nitrogen Cycle). Nitrates ions are more readily absorbed by some plants roots (e.g., potatoes, grass, etc.) absorb nitrate and ammonium ions equally well.

Nitrate of Potash. (KNO_3) Potassium nitrate. A fertilizer containing about 13% Nitrogen and Potash used particularly by horticulturists. Also called saltpetre. (Potash Nitrate).

Nitrate of Soda. ($NaNO_3$) Sodium nitrate. A fertilizer containing about 16% nitrogen derived from natural deposits of 'Caliche' in Chile (also called Chile saltpetre). A fast acting fertilizer, sometimes used as a top dressing, particularly by horticulturists.

Nitrification. The conversion by various soil Bacteria or organic nitrogenous compounds which are not available to plants, to nitrates which they can readily absorb. Organic matter from decaying animal and plant remains in the soil is reduced by enzyme action to amino compounds. These are broken down by *Pseudomonas* to ammonia and ammonium compounds, then to nitrates by *Nitrosomonas*, and eventually to nitrates by *Nitrobacter*.

Nitrochalk. A granular fertilizer containing ammonium nitrate blended with ground chalk, containing nowadays about 25% nitrogen. The chalk content reduces the acidifying effect on the soil of the ammonium nitrate.

Nitrogen. (N). An odourless, colourless, gaseous element, forming about 78% of the atmosphere. It is present in the soil in inorganic forms such as ammonium compounds (Ammonia) and nitrates, and in various organic forms such as amino acids and protein complexes as constituents of humus, which are eventually broken down by bacterial action to nitrates (Nitrogen Cycle). As a major plant nutrient nitrogen is applied to the soil in the form of Fertilizers or Manures. Nitrogen is essential to life, mainly as a constituent of protein and nucleic acids (e.g., DNA), and in plants is vital for vegetable growth. Nitrogen deficiency results in stunted plants with yellowish leaves and poor root development. Excess produces dark green succulent foliage, but if prolonged may retard the maturing of crops, and can cause cereals to lodge (Laid Crop) due to the growth of long weak straws.

Nitrogen Cycle. The circulation of nitrogen atoms and compounds in nature, mainly caused by living organisms. Inorganic forms of nitrogen (mainly nitrates) are absorbed by plants and assimilated into complex organic compounds (e.g., Proteins). Plants either die and decay, or are eaten by animals, which produce excreta and also eventually die and decay. The nitrogenous products of excretion and decay are progressively degraded in the soil, mainly by bacterial nitrification eventually to nitrates. In addition there is secondary circulation of atmospheric nitrogen which is 'fixed' and assimilated into proteins by certain bacteria living symbiotically in root nodules of various leguminous plants and by various free living bacteria in the soil (Nitrogen Fixation). A certain amount of nitrogen is released from nitrogen compounds by denitrifying bacteria.

Nitrogen Fixation. The combination by living organisms of free atmospheric nitrogen (N_2) into nitrogenous compounds is known as biological nitrogen fixation (BNF). Several types of BNF are known, classified as symbiotic (Symbiosis)or non-symbiotic. Amongst the former, aerobic bacteria (*Rhizobium spp*) living in the root nodules of leguminous plants, are probably the best known. These bacteria are able to 'fix' nitrogen, using energy supplied by the plants. A number of free-living bacteria (e.g., aerobic *Azotobacter* and anaerobic *Clostridium*) can also 'fix' nitrogen, using energy derived from soluble carbohydrates in

Humus. In addition to bacteria several other organisms are known to 'fix' nitrogen, including certain blue green algae and yeasts.

A certain amount of nitrogen in the atmosphere is combined into organic from by lighting (non-biological nitrogen fixation).

Nitrogen-free Extract. One of the fractions measured by the Proximate Analysis of Fedingstuffs, mainly containing soluble Carbohydrates such as Starch and Sugars.

Nitrosomonas. Genus of bacteria in soil and transforming ammonia into nitrous acid.

N/K Ratio. The ratio of the nitrogen content of a fertilizer to its Potash content.

Nodules. Swellings on the roots of leguminous plants (Legume) containing bacteria (*Rhizobium spp*) living symbiotically with the root tissue and capable of converting atmospheric nitrogen into nitrogenous compounds (Nitrogen Fixation) using energy derived from carbohydrates supplied by the plant. These compounds are used to assimilate protein. Some are released to the soil and converted to nitrates which may be used by other plants such as the grass in a mixed clover-grass iey. The stored protein in legume crops is released to the soil on death or ploughing in, and is degraded to nitrates (Nitrogen Cycle). Legumes such as beans or clover are thus often grown prior to cereal crops to enrich the soil with nitrogen. The various strains of bacteria which form nodules are usually naturally present in the soil, but to achieve nodulation with lucerne the seeds are usually coated or 'inoculated' with the appropriate bacterial culture before sowing. The bacteria enter the plant via the root hairs.

Normal Moisture Capacity. Amount of water which is normally held by a soil after the free or gravitational water has drained off.

Notified Kind/Variety. Seed or planting materials of any kind of variety notified under Section 5 of Seeds Act, 1966.

Noxious Weed Seed (Primary). The seeds from any plant of those species considered to be extremely destructive or harmful to agriculture. These species are designated by law for Seed Law Enforcement.

Noxious Weed Seed (Secondary). Seeds of those noxious weed species which may be present in agricultural seeds but only in limited number *i.e.*, their presence is not completely prohibited by Seed Law Enforcement.

N.P.K. An abbreviation for Nitrogen, Phosphorous, and Potassium in describing the composition of a Fertilizer. (Unit of Fertilizer, Plant Food Ratio).

Nubian. A short-haired breed of goat which originated in North-East Africa and was crossed with short-haired British goats to produce the Anglonubian breed. Characterised by long drooping ears, a Roman nose, and an arched face with prominent forehead. Various colours exist, the most common being red, black and tan, often combined with white. Noted for its milk rich in Butterfat.

Nurse-Seed Grafting. Method (asexual) of plant propagation wherein grafting operation is performed when seeds have just germinated. The petioles of the cotyledons are cut off transversely just at the seed, leaving the cotyledons inside. A knife point is inserted into the seed between the cut petioles, making an opening for the scion. Scions are prepared from the dormant wood of the previous season's growth.

Nursery. Series of plant progenies, introductions etc. grown chiefly for individual plant testing and examination in a plant breeding programme. This should not be confused with the term *nursery* used by rice growers to designate the bed where seedings are grown for transplanting to the field.

Nurse Cow. A cow used to suckle a calf produced by another cow.

Nut. 1. A small cube or cylindrical piece of compressed meal. 2. A type of fruit, dry, one-seeded, and usually with a hand woody shell, e.g., hazel nut.

Nutrient Cycling. The pathway of number substances from their occurrence in the physical environment to their incorporation in living organisms and their return to the physical environment through the metabolic activity, death, and decay of organisms.

Nutrient Deficiency Symptoms. When any essential plant nutrient is seriously lacking in the soil, plants growing on it show certain colour development in leaves and certain change in growth. These symptoms are characteristic of each nutrient. The nature of the symptom varies from crop to crop and with the degree of deficiency.

Nutritive Ratio. Ratio or proportion between the digestible protein and the non-nitrogenous nutrients including fat multiplied by 2.25, in a feed or a mixture of different feeds, Nutritive Ratio (NR) is expressed as follows :

$$NR = \frac{\text{Digestible carbohydrates} + \text{digestible fibre} + (\text{Digestible fat} \times 2.25)}{\text{Digestible crude protein}}$$

When the percentage of TDN is given the nutritive ratio is computed as follows :

$$NR = \frac{\text{TDN} - \text{Digestible protein}}{\text{Digestible protein}}$$

Thus the nutritive ratio is the ratio between the total non-nitrogenous nutrients and the digestible protein.

O

Oast. A kiln in which hops are dried. An oast-house is a building in which hops are dried and pressed into pockets, consisting of one or more kilns and a cooling room containing a press.

Oatmeal. An animal feedingstuff consisting of ground oats with the husks removed to varying extents. The husks provide a roughage element to the diet. Fed either dry or together with other constituents in a wet mash. Also fed as a porridge (sometimes called brose) to animals which are ill or recovering from sickness.

Oats. A cereal crop with a high fibre content, the grains of which are used for stock-feeding and for making bread, porridge and oatmeal. Grown mainly in the cooler and wetter areas of World, although to a decreasing extent as it yields less well than barley and is more difficult to harvest. Most cultivated oats are of the species *Avena sativa* although some *A. strigosa* is grown in the certain areas. Oats are distinguished from other Cereals by having a flowering head in the form of a tall panicle with clusters of slender spreading branches bearing Spikelets.

Objectionable Weed. It refers to those weed species, whose seed is difficult to separate once mixed with crop seed or which is poisonous or injurious or has a smothering effect on the main crop. It is difficult to eradicate once established because of its high multiplication ratio. It may serve alternate host for crop disease and pests.

Oestrogen. A sex hormone produced by female animals which controls the development of female characteristics including the Mammary Glands and controls the oestrus or 'heat', the period during which females are receptive to mating with males.

Official Seed Sample. The samples which are drawn by a seed control or seed Law Enforcement Officer. These are submitted to the laboratory to determine if a seed lot being offered for sale meets the requirements of the seed act.

Oil of Vitriol. Sulphuric Acid.

Oilseed Crops. Various crops grown for their seeds, rich in oil and extracted after crushing for the manufacture of margarine cooking oil and salad oil, etc. The residues are usually processed to produce cake, rich in protein and valued for stock feeding. Only Linseed, Oilseed Rape and Sunflower are grown in Britain. Various cakes derived from tropical oilseed crops are also commonly used (e.g., Groundnut Cake, Soyabean Cake).

Oilseed Rape. Certain varieties of Rape grown under contract for their seeds, rich in oil and used for the manufacturing of margarine, cooking and salad oils. The crop is often used as a useful break between cereals and is conveniently harvested with the same machinery. A valuable cake is a byproduct after seed crushing and oil extraction. Also known as Canola in North America.

Olericulture. Branch of horticulture which deals with cultivation of vegetables.

Olland. The aftermath following the harvesting of hay from a clover Ley.

Omasum. The third stomach of a ruminant in which food is ground following partial digestion in the second stomach or reticulum. Also called manyplies or psalterium.

Omnivore. An animal which eats both vegetable and animal foods (e.g., the pig), as distinct from a flesh-eating carnivore or a herbivore.

Open Centre. System of training fruit trees where the main stem is allowed to grow only upto a certain height by leading it within a year of planting and inducing all subsequent vegetative growth by lateral branches, resulting in a lowhead in which bulk of crop is borne to ground that in central leader tree.

Open Furrow. A type of furrow in which the furrow slices are turned in opposite directions so that they lie facing away from each other. Open furrows are used when completing the ploughing of a land or a field (the finish).

Open-Pollinated Seed. The seed produced as a result of natural pollination as opposed to hybrid seed produced as a result of controlled-pollination.

Orchard. An area of land used for growing fruit trees (e.g., apples, pears, plums, cherries, etc.) either on a small scale or on a more extensive commercial scale. Now-a-days commercial fruit varieties are normally grown on the more dwarfing rootstocks selected for their ability to control the size, growth, and cropping characteristics, since it is now economically desirable to harvest early in the life of the trees and to do so from the ground or short steps. In addition tree shape is greatly affected by the pruning system used. At the end of their useful life orchards are grubbed, the tendency being to grub trees on dwarf rootstocks at a younger age than those on vigorous rootstocks. Orchards producing desert and culinary fruit are mainly operated by specialist growers, whilst a certain number of orchards producing cider fruit are still combined with livestock grazing.

Orf. A disease mainly of sheep, but also affecting cattle and goats, due to a primary attack by a virus which causes pustules. These are secondarily infected by a fungus (*Fusiformis necrophorus*) and develop into ulcers and scabs on the lips and mouth, and on other parts including the face, legs tail, genitals, teats, etc. Also called contagious pustular dermatitis, contagious ecthyma, necrobacillosis, of and ulcerative stomatitis.

Organic. A term applied to a substance containing carbon combined with hydrogen, and very often also with nitrogen, oxygen and other chemical elements, forming complex molecules.

Organic Farming. Systems of farming which avoid the use of artificial fertilizers, pesticides or herbicides, and concentrate on methods of crop rotation and the use of home-grown feeds, organic fertilizers and manures, and ley farming.

Organic Fertilizers. Those Fertilizers derived from animal products (e.g. Dried Blood, Hoof and Horn Meal, Shoddy, etc.) and from plant residues (*e.g.*, malt culms, castor seed meal) all of which have a considerable nitrogen content. Also applied to more bulky organic materials such as Manures

Organic Matter. 1. The Organic substances in the soil, including animals and plant residues in various stages of decomposition (Humus), cells and tissues of soil organisms (*e.g.*, bacteria, fungi, insects, earthworms, etc. those organic compounds synthesised by such organisms, and organic materials added to the soil as fertilizers, e.g., Manures. The organic fraction of the soil is intimately mixed with the mineral matter. The organic matter content of soil varies according to soil type. On a dry weight basis it is usually in the range 2-15%, but certain organic soils consist almost entirely of organic matter (*e.g.*, peats, black Fen soils). 2. One of the two main components of the dry matter of a feedingstuff (the other being Mineral Matter) and consisting mainly of Carbohydrates, Fats and Proteins.

Organophosphorus Compounds. A group of synthetic non-persistent chemicals mainly used as insecticides (*e.g.* Malathion), now-a-days often in preference to the more persistent Chlorinated Hydrocarbons. They may have either a Systemic or a non-systemic action and mainly affect the nervous system.

Orpington. A general purpose breed of poultry combining both laying and table qualities. Black, buff or white in colour, with white legs and feet, a single comb, and hens laying tinted eggs.

Osier. Any willow grown as a coppice crop, the flexible twigs of which are used in basket making, particularly *Salex viminalis.* Also called withy.

Other Crop Seed. Seeds of those plants which are grown as crops, other than the main crop.

Out-Crossing. The mating of unrelated animals of the same breed, normally in order to introduce a desired character to the strain which is otherwise absent.

Outfall. The point where a drain discharges into a ditch, or where a watercourse discharges to a river or the sea (sometimes associated with a sluice or flap valve).

Outfield, Out-lying Field. 1. A field situated some distance from the farm buildings usually on the farm boundary. 2. A scottish term for a field under continuous arable cropping without being manured.

Outliers. Livestock kept outside during the winter.

Out-run Land. The open unenclosed areas surrounding a Hill Farm, normally some distance from the farm buildings.

Out-wintering. The practice of keeping cattle in the fields during the winter as distinct from keeping them under cover.

Oven Dry Soil. Soil is considered to be oven dry when it has reached equilibrium with the vapour pressure of an oven at 105°C. The oven dry condition corresponds to a relative humidity of approximately 0 per cent or a pF near 7.

Over Irrigation. Application of more water than is necessary for the needs of vegetation, resulting in loss of water through seepage and leaching with the resultant loss of humus, nitrogen and other mineral elements of the soil.

Overstocked. Condition of overcrowding in tree crops, leading to retarded growth.

Ovine Virus Abortion. Contageous disease of sheep characterized by premature birth or abortion near full term, often accompanied with retention of placenta. The disease is caused by a virus of psittacosis-lympho-granuloma-venereum group.

Oviparous. Laying eggs which hatch to release the young. Birds, amphibians, most insects and some reptiles are oviparous.

Ox. A general term for a male or female of domestic Cattle. Particularly applied to a castrated male or Steer, especially when used for draught purposes.

Oxidase. A type of Enzyme which acts as a catalyst in the oxidation of substances, removing hydrogen which then combines with oxygen to form water.

Oxidation. The combination of oxygen with a substance, or the removal of hydrogen from it. In cellular respiration, oxygen reacts with carbohydrates such as Glucose yielding carbon dioxide and water as by-products and liberating energy. This process is comparable to the oxidation of substances by burning, in which energy in liberated as heat. The term is also used to describe chemical reactions in which electrons are lost from atoms such as when ferrous iron (Fe^{2+}) is oxidised to ferric iron (Fe^{3+}).

Oxygen.(O). An odourless, colourless, gaseous element, forming about 20%, of the atmosphere and a constituent of water. Essential to most living organisms for respiration and necessary for combustion.

Oxytocin. A hormone secreted by the posterior lobe of the pituitary gland. It is responsible for activating the let-down mechanism, releasing milk from a cow's udder. It also causes the contraction of smooth muscles in the uterus.

P

Packhouse. A building in which fruit, vegetables, eggs, etc., are packed into boxes or containers for despatch from a farm to a market, retail outlet, store, etc. Casual Workers are usually employed seasonally for packhouse work.

Paddle. A long-handled, spade-like tool used to remove mud from plough shares and mouldboards, etc. In some areas called spud.

Paddock. A relatively small enclosed pasture, usually near the farm buildings.

Paddock Grazing. A system of managing grassland in which grazing areas are divided into paddocks by permanent or semi-permanent fences. Each paddock is grazed in turn and then allowed to rest.

Paddy Cleaner. Machine for removing foreign seeds, immature grains, foreign materials, such as stones, nails, dirt and dust etc. It is used prior to shelling.

Paddy Separator. Machine which separates shelled paddy from unshelled paddy and husk.

Paddy Weeder. Equipment for interculture, used in paddy cultivation. It is used for uprooting weeds and burying them in puddled soil between rows of standing paddy crop. It improves the aeration of soil.

Pale, Paling. A wooden stake driven upright into the ground for fencing. Also a wooden board nailed vertically with others to horizontal pieces to form pale fencing or paling.

Palm Kernel Cake. A cake derived from the residue of the kernels of the fruit of the African oil palm (*Elaeis guineensis*) after extraction of the oil. It has a high fibre content and is slightly gritty, and is therefore somewhat unpalatable. It is less rich in protein than other cakes and is slow to absorb water. Sometimes used to steam-up dairy cows.

Pan. A hard layer in the soil that is (*a*) compacted, such as a plough pan caused by continual ploughing to the same depth, (*b*) indurated, such as an ircn pan in which the soil particles have become cemented together by precipitated iron and silicates, or (*c*) has a very high clay content. Pans often impede drainage and root growth, and are broken down using a subsoiler. Also called hardpan or sole.

Pallets. Wooden or iron frame structures used to support a stack of bags, to prevent them from touching the floor and absorbing moisture.

Paraquat. A contact herbicide which disrupts photosynthesis causing chlorosis. Used to control a wide range of annual broad-leaved weeds and grasses. It is very poisonous to animals.

Parasite. An organism which lives in or on another (the Host) from which it derives food, for all or part of its life. Often, though not always, harmful to the host. Animal parasites include internal worms (*e.g.*, Roundworms, Tapeworm, etc.) and external Flies, Lice, Ticks, etc. Some arc parasitic on a secondary host during part of their life cycle (*e.g.*, tapeworm, Liver Fluke). Plant parasites include fungi and bacteria (causing such diseases as Mildew, Rust, Fire Blight, etc.), various plants such as dodder and mistletoe, and numerous insects and their larvae, *e.g.*, Aphids, Fruit fly, Eelworms, etc. A semi-parasite is one which feeds partly on its host but also manufacturers some of its own food (*e.g.*, mistletoe).

Parathion. An extremely poisonous Organophosphorus Compound (diethylpara-nitrophenyl-thiophosphate) used to control a variety of insect pests including Mites, Aphids, Eleworms, etc.

Parent Material. Unconsolidated, chemically weathered mineral (sometimes organic) matter derived from rock (parent rock) from which soil develops.

Parent Stock. The individual plants or animals from which others have been bred, often in large numbers.

Parkland. Park-like grassland dotted with mature, usually individual, trees.

Parthenogenesis. The growth of an ovum into a new individual without fertilization. It occurs naturally in some plant (*e.g.*, dandelion) and animals (*e.g.*, Aphids), and the offspring are normally all diploid and genetically identical to the parent.

Pasteurisation. Process of killing organisms in a product, commonly milk, by heating to a controlled temperature which is lethal for the casual agents of a number of milk-transmissible diseases without causing any material change in the natural characteristics of the substance treated.

Pasteurised Milk. Milk heated to a temperature which kills the majority of the Bacteria present, but without affecting its palatability. The treatment applied is that needed to destroy *Mycobacterium tuberculosis*, the most heat resistant pathogen normally present in cow's milk. Most milk is pasteurised using the High Temperature, Short Time method (H.T.S.T.) by which it is heated to at least 72°C(161°F) for 15 seconds and then rapidly cooled to 10°C(50°F) or less. An alternative is the Holder method by which milk is heated to between 63°-66° (140°-150°F) for at least 30 mins, and then cooled to 10°C(50°F) or less. The adequacy of the treatment and the keeping quality are determined, respectively, by two statutory tests, the phosphatase and methylene blue tests. Pasteurised milk should keep at least 2-3 days. The term is derived from the French chemist, Louis Pasteur, who pioneered the process.

Pasture. An area of grassland used only for grazing as distinct from one cut to produce Hay or Silage (a Meadow).

Pathogen. An organism or substance which causes disease.

Pears. Various high-quality varieties of pear (*Pyrus communis*), particularly Conference and Comice, are grown commercially in Orchards in the East and South-East of England. Perry pears are grown in the West and South-West of England. The area grown (except Perry pears) has declined slightly in recent years and, according to the 1980 Annual Review White Paper, was estimated at 4500 hectares (*c.* 11 100 acres) in 1979.

Peas. A leguminous crop (*Pisum sp.*) with a long trailing stem, compound leaves and leaf stalks terminating in climbing tendrils.

Peat. It consists of the remains of aquatic, bog, marsh or swamp vegetation which has been preserved under water in a partially decomposed state.

Peat Soil. Soil acidic, pH 3.9 and below, 10 to 40 per cent organic matter with partly decomposed raw plant material, suitable for paddy when water recedes.

Peck. A measure of capacity for dry goods, *e.g.*, grain, equal to 2 Gallons or a quarter of a Bushel.

Peck Order. The social hierarchy in poultry and other birds. The term derives from the habit of dominant individuals pecking submissive ones on the head.

Ped. A natural unit of soil structure as distinct from clods or soil fragments formed artificially by tillage.

Pedigree. The recorded line of ancestry of cultivated plant varieties or animals, much used in determining programmes.

Pedigree Registered Animal. One whose forebears have been recorded with a Breed Society.

Pedigree Selection. The selection of animals in stock-breeding on the basis of desirable features or qualities in their ancestors.

Pedology. The study of soils.

Peeler. A plant which has high nutrient demand and in the absence of fertilizers would impoverish the soil.

Pellet Application. Fertilizers are applied in the form of pellets one to two inches deep between the rows of paddy.

Pellet Feed. Agglomerated feeds formed by extruding individual ingredients or mixtures by compacting and forcing feed through die opening by any mechanical process.

Pelleted Seed. Layer of particulate matter around the seed. The pelleting is usually made from finally, pulverized limestone or mineral phosphate. Use of synthetic or arabic gum helps in holding the layer of these materials on the surface of the seed.

Pen. A small enclosure for animals or poultry. Also those animals kept in the pen and sufficient in number to fill it. Also called fold for sheep.

Performance Testing. The measurement of the growth rate of meat-producing animals by determining liveweight gain over a given period on a particular ration. Comparative tests between different strains or breeds are useful in the selection of breeding stock.

Permanent Wilting Percentage. The percentage water content of a soil in that condition when a crop has removed so much water from it, that which remains is insufficient to meet the requirements of the crop, as a result of which it wilts.

Permeameter. Device for measuring the permeability of soils or other material. It usually consists of two reservoirs or tanks, connected by a conduit containing the material under investigation, water being passed from one reservoir under varying conditions of head etc. through the connecting conduit.

Persistent Chemicals. Those which remain active for long periods after they are applied as herbicides, insecticides, etc.

Pest (s). Animals, insects, fungi, weed, etc., which are troublesome to crops or livestock, or cause harmful diseases, and result in considerable losses in agricultural or horticultural production in terms of both quantity and quality.

Pesticides. Various categories of poisonous chemicals used to kill Pests, including fumigants, fungicides, Herbicides, Insecticides and rodenticides. Most are synthetic organic chemicals, produced in concentrate form but diluted for application with various substances such as water, talc, clays, kerosene, etc., to ensure even distribution. They are available in a variety of forms including dusts, granules, solutions, suspensions and emulsions. Some are persistent in the environment (Chlorinated Hydrocarbons) and are dangerous to wild-life populations. A considerable amount of information about pesticides is now available to users, and the introduction and use of pesticidal chemicals is now controlled under various voluntary schemes.

Petroculture. The growing of crops for fuel. Following experiments in Brazil in the 1930s, alcohol from sugarcane and Manioc is now commonly added to petrol. Maize, Beet, and Sunflower are also under investigation in various parts of the world as fuel crops.

pH Value. A measure of the hydrogen ion concentration of a solution, and therefore an indication of its acidity or alkalinity. Expressed on a scale from 0 to 14 pH 7 is neutral, less than 7 is acid, more than 7 is alkaline.

Phagocytes. White blood corpuscles which engulf Bacteria as part of the body defence mechanism of animals.

Phenology. 1. Science which relates climate to periodic events in animals and plant life. The phenological data for crops include for example, dates of sowing, germination and emergence, dates of flowering, ripening and harvest. 2. Science that deals with the time of appearance of characteristics periodic events in the life cycle of organisms in nature, especially as those events which are influenced by environmental factors.

Phosphatase Test. A test determine the efficiency of the pasteurization of milk by detecting the presence of alkaline phosphatase which is inactivated by pasteurization. The test involves incubation of milk containing alkaline-buffered p-nitrophenyl phosphate at 37°C for a period of 2 hours. The milk is then examined by colorimetry for the presence and/or intensity of yellow colouration due to *p*-nitrophenol.

Phosphates. Various salts of phosphoric acid applied to the soil as phosphatic fertilizers to supply phosphorus to crops.

Phosphatic Fertilizers. See Fertilizers.

Phosphorite. See Rock Phosphate.

Phosphorus.(P). A chemical element occurring naturally only in compounds, mainly as calcium phosphate in the mineral, apatite. Essential to life as a constituent of certain proteins and for many metabolic (Metabolism) reactions. In animals it is important in the formation and maintenance of bones and teeth, and for efficient Carbonhydrate utilisation, and is mainly derived from Bone Flour, milk, cereals, and oil-seed Cake. In plants phosphorus is necessary for root development. It also enchances crop ripening and counteracts the weakening effect of excessive applications of nitrogen. It is applied to the soil in the form of various phosphatic fertilizers.

Photoperiodism. The effect of the length of day and night on plant flowering. Some plants are long-day plants requiring 14-16 hours of sunlight per day to flower, e.g., Wheat, lettuce, Beet, etc. Short-day plants require only 8-9 hours of sunlight to flower, e.g., kidney beans, rice, soyabean, chrysanthemum, etc. Some plants are unaffected by day length (day-neutral plants) e.g., tomato.

Photosynthesis. The process in green plants by which carbohydrates are synthesised from water and carbon dioxide using the energy of sunlight absorbed by chlorophyll. Also called carbon assimilation.

Phycomycetes. A class of fungi, mainly single-celled and found in damp situations, (*e.g.*, powdery Mildews). Those possessing thread-like hyphae lack cross walls. Many attack crops (*e.g.*, *Phytophthora infestans*, the cause of Potato Blight).

Physical Analysis. The determination of the composition of a substance by physical means. In Soil Analysis, for example, samples, are dispersed in water or simple solutions and the particles allowed to settle. After oven drying the soil is shaken through sieves of varying mesh size to separate Sand, Silt and Clay, and the proportion of each component then determined. This technique is used to accurately determine the texture of a soil sample.

Physiology. The study of the internal functioning of living organisms.

Physiological Drought. Condition when a plant is unable to take in water, because of the low ground-temperature , or because it holds substances in solution which hinder the absorption of water by the plant.

Phytopathogenic. Microorganism inciting plant disease.

Phytopathology. Science of plant disease and their control.

Picker Combine. A machine used to harvest Maize, the picking mechanism consisting of snapping rollers which remove only the cobs from the stems. 2-, 3-, or 4-row models are used.

Pickle. A grain of corn.

Pickles. Preservation of food in common salt or vinegar is called pickling. Spices and oil may also be added in pickles. The pickles are good appetisers, and to digestion and add to the palatability of the meal.

Piebald. A term applied to animals of two colours, mainly black and white, in patches.

Piecework. A basis of employment often used on farms by which seasonal workers such as fruit pickers are paid according to the amount of work done (*e.g.*, per basketful picked), as distinct from according to an hourly rate.

Pig. The pig is a domesticated ungulate derived from the wild boar kept for meat production for human consumption.

Piggery. A place where pigs are kept. A variety of forms of pig housing are in use. Breeding pigs are sometimes run outside on well-drained pastures and housed in movable sheds or arks, or in other temporary shelters. Store and fattening pigs are sometimes kept in covered or partly covered dunging yards, usually covered with straw bedding, and provided with low'kennels' as warm sleeping quarters. Modern indoor systems, particularly for farrowing pigs, usually consist of a low, insulated building provided with a heat source and controlled venetilation, and pigs are often kept in pens arranged in rows with access passages and dunging channels. Design particularly in fattening houses, varies according to whether dry or liquid feeding systems are used. In farrowing houses where Farrowing Crates are used. In farrowing houses where Farrowing Crates are used, piglets are kept warm by a covering hood housing an infrared heater.

Pightle. A small enclosure or field. Also a croft.

Pin Bones. The two main pelvic projections to either side of a cow's tail.

Pin Feather. A young undeveloped or unexpanded feather.

Pinder. A person who, in the past, officially impounded stray cattle in a parish in a pinfold or enclosure, Owners were required to pay a fine on retrieving them. Also called pounder.

Pinning. The loss of condition in sheep, and sometimes in cattle, due to cobalt deficiency. Also called pines, vinquish.

Pioneer Crop. A crop or mixture of crops (e.g., Italian ryegrass, rape and turnips) grown on ploughed-up grassland which has become worn out and the soil impoverished, prior to reseeding the grassland with a good seeds mixture. Pioneer crops are fertilized and grazed under heavy stocking (to accumulate dung and urine) and ploughed in, having the effect of improving the fertility of the soil for reseeding.

Pirls. Small knots of wool in a fleece, as found in the Wensleydale.

Pitch Pole. A type of Harrow with double-ended tines mounted on central shafts, so that they are able to be turned through 180^0. This enables clogged or obstructed ends to be turned and the 'clean' ends to be brought into play.

Placement Drill. A machine which places fertilizer close to the side of rows of seeds and slightly below them. Fertilizer placement improves the efficiency of usage by ensuring the nutrients and concentrated close to actively growing roots in young seedlings.

Plague. Infectious disease-caused by *Pasteurella pestis*. It is basically a disease of rodents which is transmitted to man and livestock. Plague is characterized by high fever, prostration, pneumonia, hemorrhage from the mucous membranes and toxemia.

Planker. An implement used to crush clods on land where a roller cannot be used, consisting of a number of fixed overlapping planks, shod with iron bars along the working edges, which is pulled over the lands. Also called float, rubber or scrubber.

Plant Breeding. Applied branch of botany dealing with the crop improvement and production of new improved crop varieties far better than original in all aspects.

Plant Consumption. Water used by vegetation in the process of growth, including that stored in the body of the plant and that dissipated from its leaf and body surfaces by transpiration.

Plant Food Ratio. The ratio of nitrogen to phosphate and potash in a fertilizer. Thus a compound fertilizer containing 20% N, 10% P, and 10% K has a ratio of 2:1:1. Such ratios are a guide to the type of compound. A number of fertilizers may have same ratio and therefore be usable for the same purpose, but the quantity needing to be applied will vary in relation to the actual percentage nutrient content.

Plant Louse. Aphid.

Plashing. Layering.

Plat. A plot of ground.

Plate Mill. A machine which coarse grinds grain, consisting of two circular plates of equal diameter positioned one above the other, with an adjustable gap between them to control the fineness of grinding. The lower plate is fixed whilst the upper one rotates against it. Plate diameter varies between 15 and 45 cm (6-18 in.)

Platyhelminthes. Flatworms.

Pleaching. Layering.

Plough. One of the oldest types of tillage implement which breaks up the land, turning over the soil into ridges and furrows, burying surface vegetation and manure and exposing the new soil surface to the air in preparation for sowing seed. The earliest types were wooden and of simple design. The principal parts of modern ploughs comprise a steel frame to which are fixed a number of bodies, each consisting of a Coulter which cuts a vertical slice, and a Share which makes a horizontal cut under the furrow slice which is then turned by the Moulboard. The lateral accuracy and stability of each body is maintained by a Landside. Three main categories of tractors-drawn ploughs are in common use on farms, viz, (*a*) fully mounted types, attached to the three-point linkage and raised or lowered hydraulically, (*b*) semi-mounted types, in which a rear wheel partially supports the weight, and (*c*) trailed types, pulled by a drawbar.

Plough Pan. Pan.

Plough-in. To bury a crop, stubble, weeds, etc., by turning them under the soil with a plough.

Ploughing. 1. Tilage operations carried out with the help of tractor drawn or bullock drawn implements known as plough, before the crops are sown. 2. Tillage operations carried out for preparing the seedbed of crops. Ploughing essentially means opening the upper crust of the soil, breaking the clods and making the soil suitable for growing seeds.

Ploughing Iron. A term for any of the iron parts of a plough, e.g., Coulter, Share, etc.

Ploughing Rate. This can be approximately calculated by the following formulae which assume a field efficiency factor of 70% to take account of headland turning and non-productive time.

$$\text{Hectares per hour} = \frac{\text{working width (m)} - \times \text{ speed (km/h)}}{14.28}$$

$$\text{Acres per hour} = \frac{\text{working width (yds)} - \times \text{ speed (m.p.h)}}{3.93}$$

Ploughland. 1. Arable land. 2. The amount of land that could be tilled in a year by a plough drawn by a team of eight oxen, together with a proportionate amount of pasture, used historially as a basis for determining taxes.

Plough Sole Placement. Method of fertilizer application in which fertilizer is placed in a continuous band on the bottom of the furrow in the process of ploughing. Each band is covered as the next furrow is turned, usually during the growing season.

Pneumatic Separator. Seed processing machine in which seed separation is made by air on the basis of differences in terminal velocity. It has a fan at the intake end and a pressure greater than the normal atmospheric pressure is builtup within the machine. The air blast is responsible for seed separation.

Plucker Finger. A wire loop in a hop picking machine which severes the cones from the Bine.

Poaching. The trampling of land when wet, mainly by cattle, so that the soil becomes churned and muddy. Poaching is a particular problem of heavy land, and leads to the deterioration of soil structure.

Pocket. 1. An elongate sack into which hops are ressed, about 2 m (6 ½ ft) tall and 60 cm (2 ft) in diameter. Also the quantity of hops contained in such a sack, about 76-89 kg (1 ½- 1 ¼ cwt). 2. A measure of wool equal to 76 kg (168 lbs).

Podsol. A type of soil in which minerals have been leached from the surface layer and deposited in the subsoil, often forming a hard pan impenetrable by roots. The upper leached layer is usually extremely poor in mineral salts and consists of bleached sand. It is often covered with very acid, decomposing plant material. Such soils are common on quick draining sandy heaths, and under conifer forest, particularly in the cool, wet areas.

Point. A share or tine.

Point of Lay. The state of a pullet when about to lay its first egg.

Points. The important features or body characteristics of animals which are considered when judging livestock.

Poke. A loosely woven sack for green hops, able to contain 8-10 Bushels (0.29-0.36 cubic m). Also called greenbag, greensack.

Pole. An old measure of length equal to 5 ½ yds (4.6 m), or of area equal to 30 ¼ sq.yds (25.3 sq.m.). Also called rod or perch.

Poll. 1. The top of the head. 2. To dehorn an animal. 3. To cut off the top of a tree.

Pollard. 1. An animal from which the horns have been removed. 2. A tree having had its crown cut off, and with new branches growing from the top of the stem.

Pollards. A term once given to a category of offal from wheat milling, consisting of fine bran and other products. All wheat offals, apart from bran, are now sold as weatings, and are used as a Feedingstuff.

Polled. A descriptive term for animals naturally lacking horns.

Pollinator. A tree planted in an Orchard to provide Pollen for the fertilization of surrounding trees. Pollinators are usually of a different variety to the rest of the trees in the orchard.

Pooking Fork. A fork bearing three tines, used to push corn swaths together into sheaves for tying.

Pore Space. The space not occupied by soil particles in the soil.

Pot Ale. A syrupy residue (Syrup Feeds) from the first distilling process of whisky which, is now being used by stock farmers as a supplement to complement low-quality roughage, such as straw, or low-quality hay. The syrup is a good source of energy and protein and also contains substantial amounts of phosphorous, potassium and copper. It has an M.E. content of dry matter higher than that of barley. Methods of feeding the syrup vary from inclusion in complete diets or pouring directly over roughage along feed troughs, to self feeding from open containers. Ball feeders are generally unsatisfactory due to the syrup's viscosity, except when the dry matter content is low.

Potash. A term applied to potassium carbonate (K_2CO_3), potassium hydroxide (KOH), potassium oxide (K_2O), and to potassium salts in general.

Potash Fertilizers. Fertilizer.

Potash Nitrate. A fertilizer containing about 15% Nitrogen and 10% Potash, derived from natural deposits in Chile, and mainly used in market gardening as a top dressing.

Potassium. (K). A metallic element found naturally in various salts and present in the soil in Clay Minerals. It is essential to life, influencing both Carbohydrate and Protein metabolism, and the movement and utilisation of water in living organisms. In plants it ensures healthy growth, improving both resistance to drought and fungal disease, and winter hardness. It assists in the transportation of the products of photosynthesis and counteracts the effects of excessive applications of nitrogen. In animals, it contributes to the maintenance of the osmotic balance of the body fluids, particularly intracellular fluid, and influences muscular and nerve activity. Continuous cropping, lack of manuring, excessive dressings of nitrogen, and certain crops with a high potassium requirement (e.g. mangolds, lucerne, potatoes, sugar-beet, etc.) may deplete the soil of potassium. It may also be deficient in sandy, peaty or chalk soils. It is applied in the form of various potash fertilizers.

Potato. A perennial herb (*Solanum tuberosum*) grown for its Carbohydrate-rich edible tuber, mainly for human consumption. Characterised by compound pinnate leaves with large and small leaflets in alternate pairs.

Potato Apple. The small green fruit of the potato, similar to a tomato.

Potato Blight. A fungal disease (*Phytopthora infestans*) of potatoes which devastated the British crop in 1845 and caused the Irish famine in 1846. Characterised initially by dark green, turning to dark brown, leaf patches, resulting in death of the haulm and loss of yield. Tubers, which become separately infected, show brown staining and usually rot.

Potato Pit. A Clamp of potatoes.

Potato Planter. A tractor-drawn machine, similar to a drill, which plants Seed Potatoes in widely spaced rows by feeding them down tubes and covering them with a ridge of soil. Various types exist including (*a*) handfed planters, in which an operator feeds the tubers into a rotating seed wheel which drops them down the tubes, (*b*) semi-automatic planters, in which the operator channels the potatoes into cups which more rapidly feed the tubes, and (*c*) automatic planters, which require no operator to feed the cups.

Potential Evapotranspiration. Evapotranspiration.

Poulard. A fattened or spayed hen

Poult. A young turkey less than 8 weeks old.

Poultry. A term usually restricted to domestic fowl or chickens kept for both egg and meet production, but also applied to ducks, geese and turkeys kept for their meat.

Most breeds of fowl are believed to be descended from the Jungle Fowl. Hybrids have largely replaced the pure-breds and crossbreds in commercial production, which is now mainly carried out in intensive, controlled-environment buildings.

Table birds (Broiler), resulting from crosses between White Cornish Game (derived from Indian Game) males and White Plymouth Rock females.

Laying birds. Most commercial ones being hybrids produced from inbred strains. Light types are derived from the White Leghorn. Medium to heavy types are often derived from the Rhode Island Red. Other breeds used in hybrid production include Rhode Island White, Barred and White Plymouth Rock, New Hampshire Red, and Light Sussex. A synthetic strain is also used, derived from several breeds. There is limited demand for pure strains of Rhode Island Red, Light Sussex and Marans.

Dual-purpose birds, which are now less popular, the Rhode Island Red cross Light Sussex being frequently used.

Fancy breeds, kept mainly for showing.

Poultry Liter. Fibrous material used on the floor of poultry houses alongwith the excreta which accumulates therein.

Poultry Science. Study of principles and practices involved in the production and marketing of poultry and poultry products. It includes breeding incubation, housing, brooding, curing diseases, feeding, marketing, management etc.

Poussin. A type of poultry killed for the table at about 7 weeks of age weighing about 0.9-1.0 kg (2-2.25 lbs) liveweight.

Power Harrow. A type of Harrow consisting of a number of p.t.o. driven transverse bars bearing Tines which reciprocate in a rotary manner. Used mainly to produce a fine tilth in a single opertion, particularly for potatoes.

Pre-emergence Herbicide. A herbicide applied following the sowing of a crop but before it emerges from the ground either before or after the weeds emerge.

Pregnancy. Condition of having a developing embryo in the body; the state of being with child.

Prepotent. A term used in breeding for a parent which has the ability to transmit more characteristics to its offspring than the other parent.

Pressed Honey. Honey obtained by pressing broodless honeycombs with or without the application of moderte heat.

Prilled Fertilizer. A fertilizer in the form of small spherical granules, about 2-3 mm in diameter.

Production Ration. Food supplied to an animal in excess of a Maintenance Ration so that it is able to gain weight, produce milk, etc.

Progeny Testing. The evaluation of the performance of the progeny of an animal (e.g., in milk production) or plant variety, used to provide a guide for its breeding value.

Progesterone. A hormone secreted by the ovaries of mammals which stimulates the uterus to prepare for pregnancy and to nourish the developing embryo. With Oestrogen it controls the development of the Mammary Glands.

Propagation. The reproduction of a species. The term is used in horticulture to mean the artificial multiplication of plants by vegetative means, including the taking of cuttings, Layering, Budding, and grafting etc.

Protectants. Materials applied to the seed to protect it from soil fungi. Certain organic mercury compounds like ethyl mercury *p*-toluene sulphamide, methyl mercury dicyan diamide, hydroxy mercuric chlorophenol etc. are highly effective protectants. Another group of fungicides, which are nonmercuric organic compounds like chloranil, thiram, ferbam, captan, zinc trichlorophenate etc. are less hazardous for operator and for the seed, but are somewhat less effective.

Protein. A class of complex organic compounds, containing Carbon, Oxygen, Nitrogen (about 16%), Hydrogen, and some also containing Phosphorus and Sulphur. They consist of thousands of Amino Acids linked in polypeptide chains and sometimes folded in a complicated manner. About 21 different amino acids are found in proteins, the sequence of their arrangement in the chains being specific for each protein and giving it its characteristic properties. Proteins may be simnple, containing only amino acids, or conjugated, being combined with other compounds *e.g.*, Carbohydrates (muco or glycoproteins), Fats (lipoproteins), nucleoproteins (phosphorus containing proteins combined with DNA and forming the Chromosomes of cell nuclei) etc. With water they form the basic constituents of Protoplasm, and play an important role in the structure of organisms. Plant proteins are often stored in seeds and other organs and include globulins, albumins, gluteins and prolamins. They are digested by animals into their component amino acids which are then used to synthesise animal proteins. Examples include Albumen in eggs, osein in bones, Collagen in connective tissue., Casein in milk globin in Haemoglobin, creatine in muscle, etc. Protein may be provided to livestock in feedingstuffs or animal origin (*e.g.*, Fish Meal, Meat And Bone Meal, etc.), or of vegetable origin (e.g., oilseed Cake, peas, beans, etc.), although the latter tend to be deficient in one or two 'essential' amino acids. Enzymes are also proteins.

Protein Efficiency Ratio (PER). It is defined as the weight gain of rate per unit intake of protein and may be calculated from the following formula:

$$\text{PER} = \frac{\text{grain in body weight (g)}}{\text{protein in take (g)}}$$

The PER value will vary with different protein sources as the composition of protein varies with regad to essential amino acids.

Protein Equivalent. An indication of the value of a feedingstuff expressed as the percentage of True Protein plus half the percentage of non-protein nitrogen, on the assumption that the latter is capable of conversion to protein by animal.

Provender. Dry food for livestock such as hay and corn. Particularly applied to cut straw or hay mixed with Meal.

Pulse Crops. Leguminous crop such as Peas and Beans yielding edible seeds, (pulse).

Pund. A sheep Fold.

Pupa. A resting stage in insect Metamorphosis between the Larva and young adult, during which movement and feeding ceases and a great change in form occurs. Also called chrysalis.

Pure Live Seed (PLS). Percentage of pure germinating seed determined by multiplying pure seed percentage by its own germination percentage and dividing the product hundred.

Pure Seed. Seed true to its kind of variety.

Pykel. A haycock.

Pyrethrins. Insecticidal compounds present in the flowers of the pyrethrum plant. The general term includes four compounds that occur together in the flowers, namely pyrethin I, pyrethin II, cincrin I, and cinerin II. The pyrethrins are effective and quick-acting against many insects and have very little toxicity warm blooded animals.

Pyrethrum. A genus of composite plants, the powdered flowerheads of which are used as an insecticide. Now less commonly used due to the introduction of synthetically produced insecticides.

Pyrophyllite. Mineral extensively used as a carrier or dilutent for insecticide and fungicide dusts. It is so named because its plate or leaflike structure (phyllon) will separated when heated (pyro). The mineral is ground to such finess that 85 per cent will pass through a 325 mesh screen. The particles have a flat or platelike structure, so they adhere quickly to plant surfaces dusted with them. The mineral looks, feels and acts like talc. But it is chemically inert or neutral to most materials used as insecticides and fungicides. It does not absorb moisture from the air.

Q

Quadrant. Frame of any shape which when placed over vegetation, defines a unit sample area within which the plants may be counted or measured.

Quality Seed. Good quality seeds are genetically true to species; or variety; have capacity for high germination, are free from diseases and insects, are free from mixture with other crop seeds, weed seeds and inert and extraneous material. The germination, and purity of the seed can be determined by conducting a seed test on a small representative sample drawn from the seed lot in question.

Quarter. 1. One of the four divisions of an Udder, including one teat. 2. A butcher's term for the limb of an animal and adjacent body parts, detached from a carcase. 3. A measure of weight, one quarter of a cwt, equal to 28 lbs. 4. A measure of capacity, equal to 8 Bushels.

Queen. Reproductive female in colonies of social hymenoptera (bees, wasps, ants etc).

Queen Excluder. A perforated metal sheet which prevents a queen bee from passing from the brood chamber of a Beehive to the honey chamber, but which allows the small workers to pass through.

Quick Hedge. A hedge of living plants, particularly hawthorn. Also called quickset.

Quicklime. Calcium Oxide.

Quickest. Quick Hedge.

Quilt. A thick, protective cover, placed over the frames of a beehive.

Quintuplets. Five offspring produced at the same birth or one of five who have been born at one birth.

Quotldlan. Recurring every day, such as malaria fever.

R

Rabbit. A burrowing rodent (*Oryctolagus cuniculus*) of the hare family.

Rabbing. Process of buring of the residues of the crops in the field after harvesting of the crop. Rabbing is followed with an idea to keep the land free from weed seeds and pest etc.

Rabi Winter Season. An important crop season of North India; refers also to the group of crops including cereals, especially wheat, fodders, rapeseed etc. Sown in October-November and harvested in March-April.

Rabies. A notifiable disease caused by a virus which infects the centralnervous system. All mammals are susceptible. It is transmitted in the saliva mainly by biting. Affected cattle, sheep, pigs and goats show abnormal changes in behaviour and develop any of the following symptoms; anxiety, aggressiveness, frequent bellowing, excessive saliva with choking, difficulty in eating, grinding of teeth and constipation or diarrhoea with staining and tail swishing. Progressive paralysis develops with death occurring on the 3rd-7th day. Also called hydrophobia.

Race. 1. A term used by breeders for the descendants of a common ancestor which exhibit common characteristics, but which differ slightly from other individuals or races within the species. Also used generally to mean a breed, variety, stock, group, class or genus of plant or animal. 2. A white streak on an animal's face. 3. A passageway along which livestock may be moved (e.g. from pen to Dipping Tank), normally made of gates set in two parallel lines sufficiently apart to take one animal at a time.

Rack Up. To fill a rack with hay or straw.

Raddle. A flexible piece or strip of wood used for weaving between uprights in the making of fences or hurdles.

Raett. A sheep fold.

Rafter. To plough land so that each furrow is overturned onto an unploughed strip.

Ragwort. Composite plants of the genus *Senecio*, particularly the coarse yellow-flowered *S. jacobaea* which is a common weed of pastures, and is poisonous to livestock. It is scheduled as an Injurious Weed and must be controlled either by cutting or treatment with weedkillers.

Rake. 1. A hand implement with a long handle attached to a crossbar bearing several prongs, used for scraping, gathering material (e.g. cut grass) together, smoothing a seed bed etc. 2. A tractor-drawn wheeled implement with long, curved, spring tines, used for gathering hay, scraping weeds into heaps, etc. 3. A very thin horse. 4. A northern term for a track, particularly one on a pasture or hillside, or in a gully. 5. A pasture. Also called raik. 6. To keep a flock of sheep moving from pasture to pasture.

Ram. An uncastrated adult male sheep. Also called a tup.

Rancidity. A chemical change in the fats resulting in unpleasant odours and taste due to formation of peroxide linkage with atmospheric oxygen attacking the double bond, thereby increasing the iodine number inspite of the fact that little glycerol and free fatty acid are released.

Rape. A plant (*Brassica napus*) related to the turnip and grown for Forage, particularly for sheep, during the autumn and early winter. There are giant and dwarf varieties which produce rapid, succulent, leafy growth, something weeds. It is sometimes included in grass Seeds Mixtures as a Cover Crop and grazed early. It is also often grown as a Catch Crop after Peas or early Potatoes, or in cereal stubble. Rape is very susceptible to Club Root and Mildew. Also called forage rape, cole or coleseed. Certain varieties possess seeds rich in oil.

Ratoon. Emergence of new crop tillers from the root in soil stocks specially in sugarcane and sorghum is known as ratoon and the practice of keeping the same crop for the next season is termed as ratooning.

Ray Fungus. A bacterium (*Actinomyces bovis*) which forms radiating threads in grasses and cereals and may cause Actinomycosis if eaten by livestock.

Reactor. An animal which is shown to possess a certain disease by a positive response or reaction to a test for the particular disease.

Readily Available Moisture. The volume of water held in the soil between Field Capacity and Permanent Wilting Percentage which is capable of being removed by plants. As a soil dries between these two reference levels the growth rate and yield of a crop are reduced.

Reap. To cut a crop, particularly a cereal, to harvest the grain.

Real Value of Seed. Value of seed sample in comparison to another defined as 'the percentage of pure germinating seeds'. It is calculated by the formula.

$$RV = \frac{\text{Purity \%} \times \text{Germination \%}}{100}$$

Reclaim. To recover land from an undesirable state and put it in a desirable state; to undertake the reclamation of land; for example, to drain wet land or irrigate arid land to make it useful for agricultural purposes.

Red Clover. A deep-rooting, short-lived, species of clover, with an erect hairy stem, untoothed leaflets, short, broad, sharp-pointed stipules, and pale rose flowers.

Red Fescue. A species of grass (*Festuea rubra*) normally under-sown in thinly sown spring cereal crops. Best suited to relatively poor upland and marginal soils.

Red Poll. A hornless, dual purpose breed of cattle derived from crosses between the now extinct Suffolk Dun breed and red cattle from Norfolk. The annual average Milk Yield was 3880 kg (3779 litres ; 830 gal.) with an average Butterfat content of 3.80%.

Red Rubies. A popular name for Devon cattle due to their deep cherry red colour.

Reddzlagg. A trade name for heat-treated calcined calcium aluminium phosphate—one of the main forms of phosphatic fertilizer now in use. It is a finely ground powder containing 32% phosphoric acid (P_2O_5) and a range of minor elements. It can be used on soils at all pH's.

Redmeat. Includes beef, lamb, mutton and venison. By contrast, whitemeat includes veal, poultry meat, rabbit and pigmeat. Pigmeat is an anomaly since it is frequently classed as redmeat by the meat trade.

Redwater. A parasitic disease of cattle caused by a microscopic protozoan, *Babesia* (*Piroplasma*) *bovis*, which attacks red blood cells releasing Haemoglobin,, and causes listlessness, salivation, fever, diarrhoea, reddish urine, and eventual death if untreated. The disease is transmitted by the common sheep tick, *Ixodes ricinus*, the intermediate Host.

Reed. 1. A tall grass (*Phragmites communis*) which grows extensively in thick reed-beds in swampy conditions. The dried stems are used in East Anglia for thatching. 2. The uterus of a Ewe.

Reforestation. The planting of trees on land where a forest has previously stood, but has been destroyed, e.g. by a forest fire. (Afforestation).

Registered Seed. Progeny of breeder's or foundation seed handled under procedures acceptable to the certifying agency to maintain satisfactory genetic purity and identity.

Relative Weed. One that may have some use to a farmer in certain circumstances.

Rendzina. A shallow type of soil developed over calcareous rocks, with a brown to black horizon of Mull Humus, often as the only Soil Horizon, over the parent rock. A neutral to alkaline soil with a variable content of free calcium carbonate.

Rennet Curd. Coagulation of milk by the action of the enzyme rennin. It is also known as sweet curd.

Rennin. An enzyme occurring in the Gastric Juice of young mammals which causes milk to clot and curdle by converting the soluble protein caseinogen into casein which then forms an insoluble calcium-casein compound. This is the principle of cheese-making and impure rennin (rennet) is extracted for this purpose from the rennet stomach or abomasum of calves.

Replacement Value (R V). It indicates extent up to which a given protein could replace some standard protein in terms of nitrogen balance. The formula is :

$$RV = 100 - \left[\frac{\text{Nitrogen balance}_1 - \text{Nitrogen balance}_2}{\text{Nitrogen intake}_1}\right] \times 100$$

$$RV = \left[\frac{\text{Difference in nitrogen balance}}{\text{Nitrogen intake}}\right] \times 100$$

Replacement values of a particular feed-stuff tells us how much quantity of protein part is equivalent to standard protein *viz.* egg or milk protein By this way we can judge the quality of protein from different sources.

Resazurin Test. A test used to determine the keeping quality of milk.

Re-seed. To re-establish a Ley by sowing seeds of grassland Species, usually in the form of a seed mixture. Reseeding may be carried out after ploughing up a previous ley which has deteriorated, or as part of a rotation

Residual Herbicide. A herbicide which, when applied to the soil, persists for some time, killing weeds on germination.

Residue. That which remains after a part is taken or separated. Agricultural residues refers specifically to the dry portion of agricultural crops as distinct from the fruit, grain or other ingradient for which the crop is grown *e.g.* stalks, straws, cobs, seed hulls, nutshells, fruit bits. Vegetable wastes, such as green pea vines and bean vines, are sometimes called vegetable residues.

Resting Pasture. One from which grazing animals have been removed to allow it to recover and put on new growth.

Restorative Crops. Crops which are grazed whilst growing so that livestock return organic matter and nutrients to the soil in the form of faeces and urine, e.g., a ley, folder roots, Kale, etc.

Reticulum. The second stomah of a ruminant which receives harder food material passed on from the first stomach or rumen and in which accidentally eaten small stones normally accumulate. Also called the honeycomb stomach after its internal mucous membrane lining which has a honeycomb-like structure.

Retting. The exposure of harvested flax to moisture so that it partially rots and the pectin, which binds the fibre bundles to the woody stem, is weakened, allowing the fibres to be more easily separated from the rest of the stem.

Reversible Plough. A popular type of plaugh having two sets of bodies, used for one-way ploughing. One set of bodies, turning furrows to the right, are used for ploughing in one direction; the other left-handed set for ploughing in the other direction. Automatic, hydraulic or mechanical mechanisms are used for turning over the sets of bodies. The reversible plough has the advantage that headland use is much reduced; no ridges, depressions or finishes are left in the field and the surface is left level. They are mainly used for deep ploughing before preparing seed beds for root and vegetable crops. Also called one-way plough.

Reversion. 1. The process by which soluble monocalcium phosphate, a constituent of superphosphate, is rapidly converted in the soil to the less soluble di-calcium salt, producing a finely divided precipitate which is easily available to plants. 2. A viral disease affecting blackcurrants causing the development of narrower and darker leaves than normal and reducing the amount of fruit produced.

Rex Rabbits. Varieties of fur breeds of rabbit with very fine, densely textured coats. They can be bred in any colour.

Rhine. A large, broad ditch or open water-course in flat marshy areas.

Rhizome. A horizontal creeping underground stem growing just below the soil surface, from which axillary buds produce new stems and roots giving rise to new separate plants, acting as a foodstore and means of vegetative reproduction, e.g. Couch Grass, raspberry.

Rhode Island Red. A hardy, medium-sized, laying breed of Fowl, rich reddish-brown in colour with black or greenish-blak tail feathers, yellow legs and a single comb. Hens lay large brown eggs.

Rice. 1. Small branches or twigs. Also pea straw. 2. A cereal (*Oryza sativa*) grown in the tropics mainly for human consumption, poor in fat, protein and minerals. Sometimes fed to poultry.

Richmond's Formula. Based on fat content by Gerber method and specific gravity at 60^0F as determined by a lactometer made for this purpose.

$TS = 0.25G + 1.2F + 0.14.$

Where T S is total solids per cent, F is fat and G lactometer degrees *i.e.*, 1000 (SG-1) at 60^0F. (Lactometer readings taken at temperatures other than 60 are corrected to by adding or subtracting 0.1 lactometer degree for each degree F above or below (60).

Rick. A stack.

Rickets. A disease of young animals due to a deficiency of either Vitamin D or Phosphorus. Bones are unable to absorb Calcium from food and fail to harden or ossify properly. The long limbs tend to bend and develop swellings on the ends at the joints.

Riddle. A large coarse sieve. Also called screen. Riddles of specified mesh size are used to sort Ware Potatoes from a potato crop.

Ridge. 1 The soil thrown up by a plough between two furrows. 2. A raised strip ploughed to mark out lands in a field before it is systematically ploughed, consisting of a balanced number of furrow slices leaning from opposite directions towards the centre.

Ridger. A type of plough bearing one or more double mould boards, each throwing two furrows simultaneously, one to each side.

Rig. A male animal in which one or both testicles have not completely descended from the abdomen to the scrotum at the normal time.

Rind Graft. Crown Graft.

Rinderpest. Cattle Plague.

Ring. 1. A metal circle fixed through the nose of an animal, e.g. in pigs to prevent grubbling in the ground with the snout, or in bulls for roping and controlling movement (also called Bulldog). Also to fix such a ring through an animal's nose. 2. A combination of buyers at a market or auction who arrange not to bid against one another to keep prices down. 3. A segment of a caterpillar or worm.

Ring Fence. A continuous fence encirling a farm or estate.

Ringbark. To remove a circular strip of Bank from a tree to prevent growth. Also called ring.

Ringbone. A callus of bony tissue which sometimes develops on the pastern bone of a horse's hoof.

Ringworm. A contagious skin disease characterised by ring-shaped patches caused by certain fungi growing on the skin surface or in hairs growing in the affected skin. In livestock young store cattle are most commonly affected.

Riparian. Land adjacent to a river bank. Also the owner of such land.

River Basin. The total area or catchment drained by a river and its tributaries.

Roan. The colour of an animal's coat, usually a mixture of white and other coloured hairs, e.g. red, brown, black, chestnut, etc.

Rock Phosphate. A natural mineral found in the form of a sedimentary rock containing various calcium phosphates. After grinding it is used as a phosphatic fertilizer. Also called phosphorite or mineral phosphate.

Roding. A term used in the fens for the practice of cutting and cleaning out dyke vegetation.

Rogue. A sub-standard plant or one of a different type, particularly in a seed crop. Also to remove such plants.

Roguing. Removal or cutting of individual plant from the seed plot which deviate in significant manner from the plants of the variety being multiplied. This is a step in the maintenance of purity in an established variety.

Roller. A tractor-drawn implement consisting of heavy rotating cylinders. Used to crush clods, consolidate soil, flatten and smooth the surface, break crusts, or prepare fields for sowing seeds.

Roller Crusher. A machine used in haymaking which passes freshly cut grass between steel or rubber faced rollers rotating at high speed, bruising the grass and allowing sap to be removed subsequently by the action of sun and air. The rollers may be cleated or grooved to provide more effective bruising.

Roller Mill. Crushing Mill.

Roman. A moderate-sized, white-coloured breed of goose, producing between 50-60 eggs per year.

Rood. An old measure of land equal to ¼ acre (1210 sq. yds) or 40 sq. Poles.

Roost. 1. A perch of resting place for a bird. Also a number of birds nesting together. 2. A henhouse.

Rooster. A domestic Cock.

Root Crops. Certain biennial plants grown for their edible swollen roots which act as food reserves, commonly in the form of sugar. The Mangel, Turnip, Swede and Fodder Beet are mainly grown as stock feed. The total area grown has decreased during this century due to high labour demands, but they are increasing again in popularity with mechanisation. Sugar Beet and Carrots and also the Potato (usually classified as a root crop) are mainly grown as cash crops. The 'tops' or foliage of root crops (except carrots and mangels) are valuable fodder, and sheep or pigs are often folded (Fold) on them. Root crops are often grown following or between cereal crops in rotation. Also called roots.

Root Graft. A graft which lies beneath the soil surface, in which the scion is joined to the root of the stock.

Rootle. To turn up earth with the snout. Also called root or rout.

Rot. Fungal or bacterial infection resulting in softing, discolouration and disintegration (in some types) of succulent plant tissues.

Rotary Cultivator. A type of tractor-mounted or trailed cultivator consisting of a horizontal shaft driven by p.t.o.from the tractor, and bearing a series of tines or blades of various designs which rotate with a 'stirring' action. Used for a variety of purposes.

Rotation. A cropping system in which two or more crops are grown in a field in a fixed sequence. If a ley is included it is known as 'alternate husbandry' or 'ley farming'. The benefits of rotation include reduced accumulation of disease and pests which accompany monoculture, weed control, the maintenance and improvement of fertility, spreading the risk of specific crop failure, and the even distribution of labour requirements over the year. In recent years farming has moved away from rigid traditional cropping programmes to more simplified systems due to various developments. These include the production of pesticides and artificial fertilizers, improved and increased mechanisation and guaranteed crop prices.

Rotational Grazing. A method of managing grassland in which successive areas are intensively grazed for a period, in which defoliation is rapid and complete, followed by a rest period during which regrowth occurs. This method (also called the on-off principle) is used in a number of grazing systems as a means of matching livestock requirements with grass growth.

Rough Grazing. A term generally applied to grassland of poor quality with a low stock-carrying capacity, particularly that covering rough land, Heath, Moorland and hill areas, where cultivation and the use of fertilizers is impractical, and which is characterised by coarse grasses, sedges, heather, bracken or mosses, etc. Such land is usually grazed by hardy hill breeds of sheep and some cattle.

Rough Stalked Meadow Grass. A close-growing perennial grass (*Poa trivialis*) with short creeping stolons, a rough, flat stem, and an erect panicled flower head, reddish, purplish or green in colour. It is common in lowland meadows and pastures, particularly on rich moist soils. It is included in Seeds Mixtures for permanent pastures, particularly those on wet heavy soils, and grows vigorously producing a low closely knit sward which remains green through the winder. It is very palatable to livestock and is also useful for hay.

Roughage. Bulky feedingstuffs containing fibre in significant amounts *e.g.*, hay and straw, which stimulate inestinal muscular activity.

Roup. A chronic respiratory disease of poultry causing loss of condition and reduced egg production. Also called croup.

Row Crops. Those crops planted in widely-spaced rows facilitating cultivations between the rows, e.g. rootcrops, potatoes, cabbages, etc.

Row-crop Tractors. Tractors designed to work in growing Row Crops. They usually have high, narrow wheels giving good ground clearance, an adjustable track width, and a narrow turning circle. They are used with special tools for cultivating between the rows. Most models are four-wheeled although some have only three wheels.

Rub. A measure of the quality of a sample of hops obtained by rubbing the sample with tho hands and estimating the amount of sticky resinous material left on the hands, and the 'feel' or silkiness of the hops.

Rubarth's Disease. Acute contagious disease affecting mainly young dogs, characterised by fever, vomitting, diarrhoea and sometimes convulsions. It is caused by a filterable virus.

Rubbed Seed. Sugar beet seeds used for sowing which have been separated from the clusters in which they are naturally fused together. Also called graded seed.

Rubber. Planker.

Rubbers. A term for Sheep Scab.

Ruddle. A colouring material (often red ochre) applied to the chest of a ram so that those ewes which have been covered (mated with) are marked. Also called keel, raddle or reddle.

Rumen. The first stomach of a Ruminant, a large sac lined with a mucous membrane in which coarse, partly chewed food is churned, partially.

Ruminants. Animals such as the cow and sheep which ruminate or in other words, chew a cud. An animals's cud consists of boluses of feed caten earlier. Ruminants have a multi-compartment stomach. By means of the microbial fermentation processes of the rumen, such animals are able to make effective use of high fibre feeds and as a result are frequently fed rations containing high levels of fibrous feeds.

Ruminats. To chew the cud.

Run. An enclosure for animals, particularly poultry, in which they are free to roam and feed.

Runt. 1. A small, undersized or stunted animal, particularly the smallest piglet in a litter. Also called anthony, cadme, deniel, dilling, dolly, sharger or wrenock. 2. An old cow or ox. 3. A Store animal. 4. A dead tree trunk or stump. 5. A cabbage stem.

Rusts. An order of small parasitic Basidiomycete fungi (Uredinales), the spores of which from spots, often rust coloured, on the leaves and stems of infected plants, *e.g.* Black Rust of barley.

Rut. A seasonal period of sexual excitement in male animals, particularly deer, when there is an urge to mate.

Rye. The hardiest of the Cereal crops (*Scale cereale).* It has a similar structure to Wheat, with single spikelets attached to each notch of the Rachis, but also has small narrow Glumes and conspicuous large outer pales tapering to long whiskery awns.

Ryegrass. One of the most important genera of grasses (*Lolium sp.*) used in Seeds Mixtures.

S

S.19. A strain of vaccine injected into heifer calves to suppress brucellosis.

Sack. 1. A measure of capacity, particularly for grain, equal to 4 Bushels. Also called coomb. Now in disuse in the grain trade since most grain is now transported in bulk by lorry. 2. A large bag of coarse material.

Saddle. 1. A rider's leather seat strapped to a horse's back. Also a similar apparatus for draught horses from which the shafts of a cart or towed implement are suspended. 2. A butcher's cut including some of the backbone and ribs.

Saddle Grafting. Method (asexual) of plant propagation. The top of the stock is cut first transversely and then two upwards cuts are given on either side to form a wedge and cleft is made in the scion so that it fits tightly over the stock, after fitting the scion, the union is sealed.

Safety Cab, Safety Frame. A structure fitted to a tractor designed to protect the driver in event of an over-turning accident.

Sainfoin. A perennial leguminous fodder plant (*Onobrychis viciifolia*), once extensively grown for both hay and silage, with Aftermath grazing, mainly in chalky areas, usually on its own, but sometimes included in seed mixtures. It has a similar nutritive value as lucerne which it is sometimes substituted for on potash-deficient soils, being somewhat drought resistant. It has a strong, woody, long root, a branching stem with pinnate leaves, pink flowers in a conical raceme, and single-seeded pods.

Salmonella. A genus of Bacteria. Many types exist some of which cause infectious diseases (Salmonellosis) in livestock. Important types are *S. dublin* and *S. typhimurium*, which often affect cattle, causing various symptoms including diarrhoea (sometimes bloody), fever, loss of appetite, lower milk yield, abortion, sometimes pneumonia, and death, etc. Infection can be transmitted in faeces, in slurry spead on grassland, in contaminated milk or animal feeds, by healthier carriers, cross infection at markets, etc. *S. typhimurium* and certain other types are the cause of food poisoning in man.

Salometer. Instrument (special hydrometer) for measuring the percentage of saturation of a brine solution with respect to salt. Used in picking and meat packing industries. Also called a salimeter or salinometer.

Salt Fertilizer. (NaCl). Common salt, sodium chloride, long used as a fertilizer for mangels, sugar beet and carrots, being applied 2-3 weeks prior to sowing, and producing increased yields.

Salt Index. Formula for ascertaining whether water is suitable for irrigation.

Salving. The protecting of sheep against cold or parasites by smearing them with a mixture, usually of tar and butter.

Sandy Soils. Soil containing a high proportion of sand particles and very little clay, usually less than 5%. Such soils have a low water holding capacity, drain freely, and due to leaching are usually poor in plant nutrients. They are also well aerated and organic matter quickly decomposes. They therefore require adequate applications of fertilizers and manures and are consequently often called 'hungry soils'. Liming is also needed to correct acidity. They tend to dry out quickly allowing them to be worked throughout the year. They are often used for market gardens using irrigation. Crops suited to sandy soils include barley, carrots, peas, potatoes, sugar beet, etc. Sometimes called 'light soils' because they are relatively easy to cultivate compared with heavy or clay soils.

Saprogenesis. Part of the life of a disease-producing organism in which it is not directly associated with a living host contrasts with pathogenesis.

Savanna. Tropica or sub-tropical grassland with scattered trees, either as individuals or in clumps. Often a transitional phase between true grassland and forest.

Sawflies. A family of insects which derive their name from the saw-like ovipositor which is used to lay eggs in twigs, leaves or blossom. The caterpillars of the apple sawfly, *Hoplocampa testudinea*, feed on the core and seed of the growing apple.

Scab. 1. A skin disease of animals characterised by scales or pustules in patches, especially one caused by mites. Also called mange. 2. A fungal disease of various kinds of fruit and vegetables, causing the growth of scaly crusts, e.g., Common Scab of potatoes (*Streptomyces scabies*) and apple scab (*Venturia in-inequalis*).

Scabies. An infectious skin disease caused by a parasitic mite (*Sarcoptes scabei*) which affects farm livestock and man and results in severe scratching. Commonly called itch.

Scalper. Seed processing machine designed to use the bulk of trash from seed so as to facilitate elevating and processing operations. It is fabricated with a vibrating or rotating screen or sieve.

Scaly Leg. A disease of poultry caused by a mite, in which hard scales develop on the unfeathered parts of the legs accompanied by irritating itchiness.

Scarification. Process of mechanically scaring or roughing the hard seed coat to make it more permeable to water.

Scarifier. A cultivator, the action of which breaks up and stirs the soil without turning it over. Also called scuffler.

Scarp Slope. A short steep slope to one side of a line of hills (*e.g.*, North Downs), as distinct from the long, gentle dip slope to the other side.

Scours. Diarrheoa in livestock, a symptom of a variety of diseases.

Scrag. 1. The neck of a sheep. 2. A lean or skinny animal.

Scrapie. A fatal, neurological disease which affects most breeds of sheep. Sheep display uncharacteristic behaviour (i.e., suddenly leaving the flock), become increasingly nervous as the disease progresses, tremble and go into convulsions. When rubbed they make chewing and nibbling movements. At a later stage rubbing against walls or posts increases and large areas of skin are exposed which become ulcerated by repeated nibbling. Death occurs after a few weeks or months of the first symptoms.

Scratch. A term for two light furrows ploughed to mark the setting out or opening of a plot, which form the base of the first two furrows (the crown furrows). Also called scribesod.

Screw. A Broken-Winded horse.

Scrub. 1. An undersized or inferior animal. 2. A stunted tree. Also an area of such trees, bushes, low shrubs and brushwood.

Scud-Winder. An old type of hedging tool.

Scur. A loose, horny knob that sometimes develops in horned cattie at the size where a horn would normally grow. Also called snag.

Scutching. The breaking up of flax stems after retting in a mill using a special turbine which separates the long and short fibre.

Scythe. A hand implement with a long wooden handle with two hand grips (nibs), bearing at its lower end, fixed at right-angles, a large gently curved blade. Used for mowing by sweeping it through the vegetation. Sometimes spelt sithe.

Seams. The angles or shaped sections of the slice turned over in ploughing where they touch each other.

Season. 1. One of the four climatic divisions of the year, *e.g.*, spring. 2. The oestrus or 'heat' period of a female animal. 3. To prepare a seed bed.

Second. To hoe between row of rootcrops which have previously been subjected to singling.

Scrutch. To rub the rotted stems of flax, hemp, or other fibre crop with a comblike instrument so that the stems are broken and the fibres laid bare for further processing.

Second Cut. The second crop or cut of grass taken from a field in a season for hay or silage.

Seconds. Those grains of a cereal crop intermediate in size between the largest ones (head corn) and the smallest ones (tail corn) threshed out during harvesting with a combine harvester.

Sedges. A large family of grass-like plants (*Cyperaceae*), particularly the genus *Carex*. They are distinguished from grasses by having a solid, usually triangular, stem, and one scale beneath each flower (grasses have two).Flowering heads are greenish, brownish or purplish. Sedges are common in wet or badly drained areas and are regarded as weeds.

Seed. 1. A reproductive structure of flowering plants, conifers and various other plants. It develops from a fertilized ovule and comprises an embryo and a food reserve contained in a protective coat or testa. 2. To sow seeds of a crop in a field.

Seed Barrow. A hand-pushed or tractor-mounted implement used for sowing seeds, consisting of a long narrow box with a perforated base, mounted laterally on a wheel barraw-like frame with a single wheel. The seeds were brushed through the holes. Also called shandy barrow.

Seed Bed. An area of land or a field cultivated to a level, fine tilth, in which seeds are sown.

Seed Blowers. Mechanical device by which light weight is removed by blowing.

Seed Borers. Many insects bore into seeds, either to feed on them or to lay their eggs in them. Some attack fresh growing seeds: other dried seeds such as stored grain. Chalcid wasps bore holes in seeds with their long ovipositors and deposit eggs inside larvae already present within the seed. Codling Moths such as *Laspeyresia pomonella* by there eggs on leaves and flowers of apple tree and the newly hatched eaterpillars bore into the developing seeds; weevils such as the Grain weevil and the Bean weevil use their snouts for boring into dried seeds.

Seed-borne Diseases. Those plant diseases which may be transmitted from one generation to the next, either within or on the surface of seeds, e.g., many fungal diseases.

Seed Certification. Maintaining and making available to the public, the seeds of high quality and superior propagating materials so as to ensure genetic identity. This is achieved through continuous inspections of fields and seeds and by regulations for checking on the production, harvesting and cleaning of each lot of seed.

Seed Cleaning. The removal from seed, usually by machine, of impurities such as weed seeds, broken leaves, insects, etc.

Seed Cum Fertilizer Drill. Seed drill fitted with fertilizer dropping attachment, distributes the fertilizer uniformly on the ground.

Seed Corn. A cereal grown to provide grain for seed.

Seed Disinfection Ridding the seed surfaces of organisms which are potentially disease producing.

Seed Dressing. The chemical treatment of seeds, particularly cereals, with fungicides and sometimes insecticides, to protect them against soil and seed-borne diseases and pests. A dye may be added which enables treated seed to be distinguished.

Seed Drill. A tractor-mounted or trailed machine which sows seeds in rows. It consists of a seed hopper incorporating seed metering units which are generally driven by gearing or chains from the land wheels. These units, which are adjustable to suit various seed size and sowing rates, transfer the seed from the hopper down coulter tubes to a groove cut in the soil by coulters.

Drills for cereals and grasses normally sow the seed at random, while precision drills used mainly for sugar beet and vegetable crops, place individual seeds at specific pre-set intervals in the rows. The feeding mechanisms that are available include :

Random Spacing Drills	*Precision Drills*
Internal force feed	Belt feed
External force feed (fluted or studded roller)	Cell-wheel feed Pneumatic or vacuum feed
Brush feed	Cup feed
Sponge roller feed	

Seed House. A commercial company providing seed varieties for crop production.

Seed Inoculation. The dressing of seeds of leguminous plants with a culture of nitrogen-fixing bacteria, usually when the soil is deficient in the bacterial strain appropriate to the legume concerned and would otherwise restrict root nodule formation.

Seed Leaf. A cotyledon.

Seed Potato. Relatively small tubers, usually 32-57 mm (1 ¼ 2 ¼)in diameter. Produced from approved parent material under the statutory certification scheme administered by the respective agricultural department the countries. Growing crops of 'seed' are officially inspected for health, purity and vigour prior to certification. Two categories of 'seed' are certified, viz, (*a*) Basic Seed, used mainly for multiplying further as 'seed', and (*b*) Certified Seed, used for the production of crops of Ware Potatoes.

Under the Seed Potato Regulations, grading is within the following categories, some of which have specified levels of tolerance of disease, particularly viruses.

Basic seed:	V.T.S.C. 1 V.T.S.C. 2	Virus Tested Steam Cuttings
	S.E. 1 S.E. 2 S.E. 3	Super Elite Grades
	E.	Elite
	A.A.	First Quality Commercial Seed
Certified Seed	C.C.	Healthy Commercial Seed

Seed Quality. The quality of crop seeds has two principal components. Firstly, the value of different varieties for crop production, assessed by yield trials and tests for disease reaction and produce quality. Secondly, the value of different 'seed lots' for sowing assessed by laboratory analysis of samples for purity, germination, weed content and seed-borne diseases.

Seed Royalties. Fees paid under licence for the use of seed of protected (patented) plant varieties, and usually included in or added to the sale price of seed. Royalties are paid to breeders and are used to finance continued plant breeding work.

Seed Sample Certificate. Form of international seed Analysis Certificate (Blue Certificate) use when sampling from the seed lot is not under the responsibility of a member station.

Seed Technology. Science which studies the production, harvesting, processing testing, packaging, storing and marketing of seeds.

Seed Testing. It provides information to meet legal standards, determines seed quality, and establishes the rate of sowing for a given stand of seedling.

Seeding Year. The year in which a grass (Seeds Mixture) is sown.

Seedless Hay. Hay derived from the remains of a grass seed crop after threshing out the seed heads with a combine harvester.

Seedless Hops. Hops, the cones of which lack seeds, not having been pollinated.

Seedling. A young plant, grown from seed as distinct from one grown from a cutting or by a graft.

Seedling Rootstock. A stock grown from a pip or seed of a wild crab apple. Also called seedling crab.

Seed-lip. A basket carried against the side of the body and supported by a strap round the neck, from which seed is sometimes broadcast by hand. Also called seed cord or sidlip.

Seeds. A lay usually a mixture of grasses and clover.

Seeds Harrow. A light type of Zig-Zag Harrow, with short, straight tines, usually comprising several sections attached to a frame. Used to prepare seeds beds and over seeds after sowing, and often hitched behind the drill. Sometime also used for light weeding.

Seeds Lay. A short-term ley, sown with a seed misture of grasses and clover.

Seeds Mixtures. Mixtures mainly of grass and clover seeds, sometimes including other herbage legumes, sown to produce both short-term and long-term Less. Mixtures are precisely compounded with varieties suitable to the conditions and needs for which the ley is required. They may be prepared to order by a merchant or standard mixtures can be purchased from seeds firms. Also called small seeds.

Seedy-cut. A blackish blemish sometimes found in pigmeat.

Seepage. The emergence of Groundwater from the soil surface in an 'oozing' manner, often along an extensive line, as distinct from a continuous flow from a spring at a particular spot.

Segmented Seed. The 'seeds' of certain rootcrops (e.g., Mangel, Sugar Beet) naturally borne in clusters of fruit each containing a single true seed, which are often separated for sowing. If the cluster itself is sown, several seedlings may develop which makes singling a slow operation.

Selection. The basis of animal and plant breeding in which individuals are selected for breeding on the basis of specific desired characteristic or qualities.

Selective Cutting (Forestry). System of cutting in which single trees, usually the largest, or small groups of such trees, are removed and reproduction secured under the remaining stand and in the openings.

Selective Herbicide. A herbicide capable of killing or stunting the growth of weeds growing in a crop but having little or no harmful effect on the crop itself.

Self-feed Silage. A management system whereby stock are allowed to graze silage *in situ*, the amount taken being controlled by an electric fence or movable barrier set a short distance from the silage.

Self-mulching Soil. A soil, the surface of which is so well aggregated that it does not crust over under rainy conditions and is able to act as a surface mulch when the soil dries.

Self Pruning. Natural death and fall of branches, especially the lower branches, from the live plants due to causes such as light and food deficiencies, decay, insect attack, snow and ice.

Self-sown. A term applied to a plant which grows from wild as distinct from sown seed.

Self-sterile Plant. A plant which is not able to fertilize itself with its own pollen.

Semen. A liquid produced by the male reproductive organs of animals containing seprm.

Semi-digger. A type of mouldboard with a gently concave curvature and twisted along its length, sometimes with a renewable leading edge or shin, and producing a broken furrow slice. It is in general use for various operations including moderately deep ploughing (*e.g.*, for rootcrops), ploughing-in farmyard manure and surface trash, rapid seed bed preparation in the spring, and winter ploughing on well drained soils. It ploughs to a depth of 30-35 cm (12-14 in.) and is therefore intermediate between the digger and general purpose mould-board.

Semolina. Particles of fine, hard wheat which do not pass into flour during milling.

Sepal. One of the green leaf-like parts of a flower which form the calyx and surround the petals.

Separated Milk. Milk from which the cream has been removed by centrifuging. It has about half the energy value of whole milk, but a relatively higher content of protein and minerals and is used as an animal feedingstuff.

It has a dry matter content of about 9%. The removal of fat means that it is deficient in the fat-soluble vitamins. A, D and E, but it is a good source of the B complex vitamins. It is sometimes fed to pigs in the liquid form combined with barley meal. More usually it is reduced to a dried powder, with a protein content exceeding 30%. Such powder has various uses such as the production of substitute milks for calves and orphan lambs (when vegetable oils are incorporated before drying), and inclusion in chick rations and in meals for creep feeding to piglets. Also called skim milk.

Serve. To copulate with. Also called cover.

Service. The act of a male animal serving, covering or copulating with a female.

Service Sample. Seed sample submitted to the Central Seed Laboratory, or a State Seed Laboratory and the results of which are to be used as information for seedling, seeling or labelling purposes.

Set. 1. To form fruit or seed. 2. To put a hen on eggs to hatch them or to put eggs under a broody hen. 3. To plant young plants, bulbs and potatoes. 4. A seed potato or a piece of a potato used for planting. 5. A plant cutting or shoot used as a scion in a Graft, or placed in a rooting medium or the grow roots. 6. A badger's burrow. 7. A breeding group of geese, usually comprising 1 gander with 3-5 geese. 8. A young hop plant. 9. To adjust a plough for work.

Set-to. An orphan lamb fostered by another ewe.

Set-Up. To mark out the headland in a field with a market furrow, and the various lands with ridges, before commencing ploughing.

Sewage Sludge. A partially dried residue from sewage treatment works sometimes used as a fertilizer. Compared with Farmyard manure it is poor in potash and less beneficial on soil texture.

Shab. A term for sheep scab.

Shamble. A slaughterhouse or a meat market.

Share. A pointed steel or cast iron blade attached to the front of a mouldboard on the Body of a plough which makes a horizontal cut under the furrow slice which is then turned by the mouldboard. Also called plughshare or point.

Sharp.1. A meal in which the particles are clearly visible. 2. A descriptive term for light, gravelly land which tends to wear implements.

Sharps. A term once given to a category of wheat offal resulting from milling. All wheat offals, apart from bran, are now sold as weatings for use in feedingstuffs.

Shaw. 1. A small wood or coppice. 2. The aerial parts of rootcrops.

Shear.1. To cut or clip a fleece from a sheep. 2. A term used in loosely describing a sheep's age. For instance, a three-shear ewe is between 3 and 4 years old, having been shorn three times.

Shearing. The clipping of a fleece from a sheep, now-a-days almost always by machine shearing as distinct from the old practice of hand shearing. This normally takes place when the warm summer weather has caused the yolk (grease) to rise in the wool.

Shearling. A young sheep between its first and second shearing.

Shed. 1. To separate one or more animals from a flock or herd. 2. Cereal crops are said to shed grain when a strong wind or heavy rain, etc., causes the grain to drop from the ears.

Shedder. A race leading to a two-way swing gate, along which sheep may be driven in order to separate or shed some individuals from a flock.

Sheep Cote. An enclosure for sheep.

Sheep Dip. An approved chemical used in a dipping bath diluted in water, to disinfect sheep to control parasitic diseases such as sheep scab.

Sheep Maggot Fly. A type of Blow Fly (*Lucilia sericata),* also known as green bottle, metallic blue-green in colour and of similar appearance to the house-fly. It lays its eggs in wounds and in the fleece, and on hatching the maggots bore into the flesh, causing the condition known as strike.

Sheep Names. The names given to sheep are probably more numerous than for any other of the domesticated animals, and a great many local terms and variations exist.

Sheep Nostril Fly. A type of Bot Fly (*Oestrus ovis),* greyish yellow-brown in colour, which either lays its eggs or deposits the hatched maggots in the nostrils of sheep. The maggot crawls upwards towards the sinuses causing a nasal discharge and the condition known as staggers. When fully grown it is discharged by sneezing and forms a Pupa in the ground.

Sheep Pox. A very contagious viral disease (*Variola ovina)* of sheep causing fever, cessation of feeding, difficulty in breathing, depression and skin eruptions, with a severe loss of condition if badly affected. In Britain it is a notifiable disease with compulsory slaughter of all affected sheep.

Sheep Pulmonary Adenomatosis. A highly contagious, progressive pneumonia affecting the lungs of sheep.

Sheep Run. An area of open grassland used for keeping sheep on.

Skeep Scab. A notifiable disease of sheep caused by the mite *Psoroptes communits,* which lives on the skin surface and feeds on Serum emanating from puncture wounds which it inflicts and which become inflamed, forming scabs. An irritant poison secreted by the mites causes the sheep to scratch and rub itself, with the fleece becoming detached in patches revealing the scabs, which are themselves rubbed off leaving ulcerations. Prevention is by compulsory dripping.

Sheep Walk. A range of pasture on which sheep are allowed to graze.

Sheep-gate.1. A hurdle used for penning sheep. 2. A right of graze sheep on land. Also the land on which such a right exists.

Sheet Erosion. The erosion or removal of surface soil in a uniform manner under the influence of surface run-off water.

Shell. To remove the grain from the ears of cereals.

Shelter Belt. A line or 'belt' of trees purposely planted to act as a shield and provide shelter against the weather, particularly the prevailing winds. Soil erosion is checked and crop yields show an improvement.

Shift. One of the crops in a Rotation (*e.g.,* the barley shift), or the entire sequence or rotation itself (*e.g.,* a four-course shift).

Shim. A horse shoe.

Shin. The leading edge of a mouldboard which in some types (*e.g.* Digger, Semi-Digger) may be replaced when it becomes worn. Sometimes called a cutter.

Shire. The largest and heaviest breed of draught horse, with well feathered legs, a heavy head, short arched neck, and a shortish, wide, very strong, muscular body. Males can stand over 17 hands high and may weigh over 1000 kg (1 ton). Variously coloured, often with white markings on the head and feet.

Shoddy. A waste product of the woolen industry consisting of discarded shreds and fragments of material, used as a fertilizer, particularly by horticulturalists. Its quality depends on the amount of pure wool (which is entirely protein) in it. The presence of other waste products (e.g. cotton) results in the nitrogen content varying considerably in the range 3-12%.

Shoot the Red. The stage in growth of a young turkey when the face reddens and the wattle develops.

Short Manure. Another term for short dung.

Short Sheep. Short-woolled sheep.

Short Work. The ploughing of those odd corners left unploughed (short land) during the main systematic ploughing of a field.

Shots. Small lambs which are sometimes culled.

Shut Up. To prevent stock from having access to a meadow or pasture so that the grass is able to grow and subsequently be cut for hay or silage.

Siblings. Progeny with the same parents, lie., brothers and sisters.

Sick. Diseased. Sick land (sometimes called stale land) is land which has been continuously used in the same manner over a long period of years (e.g., barley grown year after year, or grazed by the same stock over many years) so that pests and diseases have built up in the soil. Examples include fowl sick, pig sick and sheep sick land, etc. Such land should be put down to another crop or rested from the particular stock. Plough-sick land is and which needs sowing to a ley.

Sickle. 1. A hand implement with a short handle and a hooked-shaped, toothed blade, used mainly for cutting cereals. 2. One of the long, curved feathers of a cock's tail.

Side. A butcher's term for a half of a carcase, divided along the backbone.

Side Graft. A type of graft in which the scion is attached to the side of a branch rather than into the cut end of a stock.

Sideland. A strip of land left at the side of a field to be ploughed out with headlands.

Sideling. Land which slopes sideways in relation to the line of work, e.g., when ploughing along the contour.

Sieve. 1. A frame containing a mesh or perforated base for sifting material. Normally finer than a Riddle. 2. A basket for fruit with a bushel capacity.

Silage. A feedingstuff consisting of forage crops (e.g., Grassland Species, kale, beet tops, pea haulm, etc.) cut or harvested in the green state and preserved in a silo in a succulert condition for later use. The principle of silage making is the fermentation, by bacteria, of carbohydrates in the plant material to organic acids, and of Proteins to amino acids, which act as preservatives. Well made silage is yellowish-brown in colour and contains mainly lactic aid (as much as 20% of its fresh weight) produced by *Lactobacilli*, with some acetic acid and with a pH or 4.0 or less. Crops contaminated with faeces and soil tend to produce a dull olive green silage containing undesirable butyric acid, produced by *Chlostridia Lactobacilli* grows best when the silage material has a high sugar content. Crops are ideally cut after dry sunny weather which stimulates photosynthesis, and therefore sugar production, and wilts the plant, increasing sugar concentration. Cut crops may also be crimped or lacerated to increase wilting.

When sugar content is low various additives are sometimes included, e.g., molasses, which provides supplementary sugar and formic acid, which increases acidity and suppresses *Chlostridia*.

Silk. The styles or 'tassels' of the female flower of maize.

Silo. A container in which silage is made and stored. A number of types are in use including the various forms of clamp on the surface, in a pit, or in a large sealed polythene 'bag' evacuated of air. Silage may also be prepared in wooden, concrete or steel towers (sometimes under airtight conditions) in which carbon dioxide production restricts respiration and allows good fermentation. The term silo is sometimes applied to a moist grain storage tower.

Single. To reduce the number of plants in a rootcrop row by separating and removing all except one of the individual plants in each clump.

Single Comb. A type of narrow, serrated, erect Comb in poultry.

Single-suckling. A method of feeding beef cattle in which calves are allowed to suckle their own mothers.

Singletons. Single lambs born to Ewes, as distinct from twins, triplets, etc.

Sire. The father of an animal. Also to be get offspring as the father.

Sisal. A species of plant, *Agave sisalana*, native to Mexico and Central America, grown for its hard fibre which is used for manufacturing binder twine and rope.

Sit, Sitter. To Brood, a Broody Hen.

Six-tooth Sheep. Sheep of about 24-27 months of age.

Skep. An old type of straw of wicker beehive.

Skewbald. Coloured white and another colour (except black) in irregular patches.

Skim Milk. Separated Milk.

Slack Bine. Hop bine with relatively few branches and a shortage of foliage.

Slack Dried Hops. Insufficiently cured hops, with a moisture content exceeding about 12%.

Slaughterhouse. A place where animals are hygienically slaughtered and carcases are prepared for sale to the public for human consumption. All carcases are examined by qualified inspectors employed by the Local Authority and only those free from disease are allowed to leave the premises for sale to the public.

Slender Foxtail. An annual grass commonly found on heavy and poorly drained land. An arable weed, particularly in early sown winter cereal crops. A tufted, smooth-sheathed, green or purplish grass, with a slender, cylindrical, tapering panicle. Also called black grass.

Slice. The strip of soil turned over in ploughing to leave a furrow. Also called sod.

Slink, Slink Calf. A prematurely born animal, particularly a calf; or a very young calf, or one removed from its mother's womb when the latter has been slaughtered. The meat, sometimes called slink veal, is used mainly for such products as meat paste.

Slip. 1. To miscarry or abort. 2. A scion or cutting.

Slob. Land in a muddy condition.

Slot Seeding. A method of re-seeding pasture without destroying all existing turf. A special drill was produced which opens up slots about 25 mm (1 in.) wide and deep, sprays a band of herbicide to kill off adjacent competition and sows the grass seeds into the base of the slot. Other machines have now been developed with multiple sets of rotary cultivators and seed spouts which open and seed larger slots. The rotary action copes better with thick mats of old grass and roots, and the floating action of the individual units copes with uneven ground.

Slough. A muddy hollow or a marsh.

Slubbing Out. A term used in the fens for cleaning dykes.

Sluice. A structure in a watercourse used to control the flow of water, usually by means of a door, flap or hatch. Also a drainage channel.

Slurry. A semi-fluid mixture of faeces and urine, often also containing rain water and washing-down water from livestock buildings. It is sometimes mixed with litter, mainly straw, to produce farmyard manure. It may also be stored in a lagoon or tank, where it is diluted with water and is then piped onto fields (when fibre-free), or it may be sucked, under vacuum, direct into a tanker and directly distributed on fields. Machines are also available which pass the slurry through a hydraulic press providing a 'solid' friable residue which can be composted, the liquid being spread on the land. Slurry contains valuable nitrogen, phosphate and potash but its composition varies according to the type of livestock, diet, dilution, etc.

Smallholding. A general term for a small farm.

Smedi. An acronym for a complex of viruses which are associated with Stillbirths, Mummified Foetuses, Embryonic Death and subsequent Infertility in pigs.

Smooth Stalked Meadow Grass. A tufted, rhizomatous, perennial grass, with smooth, greyish-green leaves, with either an abruptly pointed or a blunt hooded tip and a smooth-sheathed, folded stem. The purplish, green or greyish flowers are in the form of an ovate panicle. It is an important hay and pasture grass in North America and in continental Europe.

Smother Crop. A crop which grows vigorously when well fertilized (e.g. Kale, arable silage crops, etc.) and occupies most of the growing space, tending to retard or smother the growth of weeds.

Smout. A hole in a wall through which only one sheep at a time may pass. Also called lunky.

Smudging. The practice of making smoke to produce an artificial cloud, in an attempt to reduce heat loss from the ground and revent frost damage to crops.

Smuts. An order of Basidiomycete fungi which produce black spores many species causing plant diseases. The main species of agricultural importance attack cereals, the spores developing in, the grains which are destroyed, e.g. Bunt.

Snath, Sned, Sneath. The curved wooden handle or shaft of a Scythe. Also called snead.

Sned. 1. To ut branches from trees or to prune. 2. See Snath.

Sock. A plough share.

Sod. A slice.

Sodium. **(Na).** A metallic element, an essential constituent of the body fluids in animals. It is mainly present as various salts, mainly sodium chloride (NaCl) or common salt, its concentration in the blood being finely controlled with excesses excreated. Reduced performance results from sodium deficiency and livestock diets are usually supplemented with salt in mineral mixtures. Sodium chloride is also used as a salt fertilizer.

Soft Fruit. Loosely berried Fruit with relatively soft flesh, divided into bush fruit (e.g. black, red and white currants, goose-berries, etc.) and cane fruit (e.g. raspberries, blackberries and hybrid berries), and also strawberries. The latter are the most widely grown soft fruit, the principal maincrop areas being in Kent and East Anglia. Blackcurrants are also widely grown in Britain, mainly for the preparation of juice for the manufacture of soft drinks and flavours for confectionery. The largest concentration of raspberry plantations in the world is located around Perth in Scotland.

Soil. The unconsolidated material covering the surface of the earth in which plants grow, and in which many animals (e.g.insects, worms, beetles, bacteria, etc.) live and derived food. It consists mainly of particles of Sand, Silt and Clay, closely associated with Organic Matter, the relative proportions of each determining the soil type. Soil mineral matter is derived from the weathering and erosion of rock. Water percolating through the soil depletes the surface layers of both slouble and fine insoluble substances, and has an effect on the hydrogen and hydroxzyl ion content of the soil, thus determining its acidity. From the agriculture viewpoint the top horizons, which crop roots penetrate to obtain water and nutrients, are the most important.

Soil Air. The gaseous content of the soil (i.e., that part of the volume of the soil other than solids or liquid) occupying part of the capillary spaces between soil particles and the soil pores. Soil air differs from atmospheric air, having a higher concentration of carbon dioxide due to respiration by plant roots and micro-organisms, and is normally saturated with water vapour. The carbon dioxide dissolves in the soil moisture forming carbonic acid, which assists in dissolving mineral matter to release plant nutrients.

Soil Analysis. The determination of the composition of soil by various laboratory chemical and physical methods. Mechanical analysis is used to accurately determine soil texture, an important factor in soil and land use classification. The analysis of soil in terms of its chemical composition is important in revealing its state of fertility and in recommending the amount and type of fertilizer treatment required.

Soil Classification. The systematic arrangement of soils into various categories according to their characteristics. This may be (*a*) according to texture, e.g., sandy loams, clays, etc., (*b*) on a geological basis, e.g., chalk soils, sandstone soils, etc., or (*c*) on a pedological basis. The basic pedological unit of classification now used for soil mapping is the soil series. The five most important soil groups are brown earths, calcareous soils (e.g., rendzinas, brown calcareous soils), Gley soils, Peat soils and podsols.

Soil Horizon. A layer of soil approximately parallel to the surface with fairly distinctive properties such as texture, colour, structure, mineral or chemical composition, etc. A vertical section of soil (a soil profile) will reveal the various horizons present from the surface to the parent material (if soil development has been *in situ*).

Soiling Crops. Crops harvested when green and succulent condition and fed to animals shortly after cutting. It is neither dried nor stored for future feeding.

Soil Moisture Deficit. The amount of water which a soil requires to be added to return it to field capacity. The size of this deficit, together with the condition of the crop, provide an indication of whether irrigation is required.

Soil Scoop. Equipment used in excavating ditches, clearing drains and moving soil over short distances.

Soil Series. The basic unit of soil classification and mapping consisting of soils which are identical in all the main profile characteristics, except surface texture, and predominantly confined to one parent material. The soil series can be considered as comparable to the genus in the classification of living organisms. It can be subdivided into lower categories (comparable to Species) on surface texture, depth, slope, etc.

Soil Sterilant. A chemical applied to the soil to kill pests, diseases or weeds.

Soil Texture. 1. The 'feel' of a soil, e.g., soapy, gritty, sticky, etc. Soils containing small clay particles, with small interparticle spaces are called close-textured, as distinct from open-textured soils containing larger particles (e.g., Sand) with larger air spaces. 2. A soil sample may be describe as belonging to a particle class of soil texture (e.g., sandy loam, silty clay, sandy clay loam, etc.) following Physical Analysis to determine the relative proportions of clay, sand and silt. The various textural classes have strict compositional limits, e.g. sandy clay loam containing 20-35% clay, less than 28% silt, and 45% or more sand.

Soilage. Fodder crops, cut and fed to livestock in the fresh state. Also called zero grazing.

Soiling Crop. A green crop cut and fed to stock elsewhere, i.e., not at the field where grown. Such cut forage is sometimes called green meat.

Sole. 1. The base of the body of a plough which rests in the furrow bottom. Also called slade. 2. See Pan.

Sole Crop. Sowing of only one crop in same piece of land in one season *i.e.,* entire crop where, there is no mix crop *e.g.,* growing of sugarcane, potato etc.

Sole Furrow. The last slice cut during ploughing. Also called scolt.

Solids-not-Fat. The various constituents of Milk other than butterfat and water. On average they represent about 8.7% of milk and consist of Proteins and other nitrogenous substances. Lactose or milk sugar, and various mineral salts and vitamins, etc.

Soot. Soot from domestic chimneys and boilers containing ammonium sulphate is sometimes used as a nitrogenous fertilizer.

Sore Mouth of Cattle and Horse. Infectious disease of cattle swine and horses, characterized by fever and formation of vesicles on the buccal mucosa and certain parts of hairless skin such as coronary bands, interdigital space, teats and udder. It is caused by virus having rod like particle 310 mμ long and 60 mμ in diameter.

Sore Muzzle of Sheep. Virus disease of ruminant (mainly in sheep), characterized by hyperemia and ulcerative inflammation estinal tract. It is caused by a virus of particle size 100-150 mμ.

Sorghum. A tropical cereal (*Sorghum sp.*) the grain of which is imported for feeding to poultry.

Sour Soil. An infertile cid soil, as distinct from a sweet or fertile alkaline soil.

Sow. 1. An adult female pig, after having produced her first litter. 2. To scatter seed on the ground by broadcasting or to place it in the ground, usually with a drill.

Sowing Behind the Plough. Very common method used for seeds like maize, gram, peas, wheat and barley. A man drops seeds in the furrows behind the plough.

Sowing, Broadcast. Scattering of seed more or less evenly over a whole area, either one on which a crop is to be raised directly or a nursery bed, as opposed to other forms of sowing such as patch or strip sowing.

Soyabean. A leguminous plant grown in warm temperate climates producing beans rich in oil and protein. Soyabean Cake and Meal are imported mainly from Canada and the U.S.A. and are used to balance the cereal content of livestock rations. They are a particularly rich source of the amino acid lysine.

Spacing Drill. Another term for a precision seed drill.

Spade Husbandry. The cultivation of land with a spade rather than a plough.

Spado. A castrated or impotent animal.

'Spaghetti' Wheat. A term sometimes applied to a variety of wheat used for the production of spaghetti and other pasta products.

Span. A team of oxen or a pair of horses.

Spangled. A term for poultry feathers with a shiny dark spot at their tips.

Spay. To remove the ovaries of a female animal to prevent it from breeding.

Spayn. To wean.

Spean. A teat.

Speckled Yellows. A disease of sugar beet due to manganese deficiency, in which the leaves become yellow between the veins, giving a streaked appearance. Some parts may turn brown. The disease is common on soils with a high lime content.

Spelt. An inferior species of wheat in which the inner husks stick to the grain, as in oats. Grown in mountain areas in Europe.

Sperm. A male gamete. They are produced by male animals in millions in a liquid medium called semen.

Spile. A cleft piece of wood used for fencing, either linked by wire or nailed to a horizontal batten.

Spine. 1. The backbone of an animal. Also called spinal column. 2. A stiff thorn or part of a plant, often a modified branch or leaf. 3. The heart wood of a tree or other woody plant.

Spinner. A tractor-drawn implement used for harvesting potatoes. Trailed and mounted types are in use. The main components are a digging share which lifts the row of potatoes and loosens the soil, and a revolving wheel set at right-angles to the row and bearing rotating tines. The tines strike the lifted row finging the potatoes and soil to the side often against a ropenet, so that much of the soil passes through and the potatoes fall to the ground in a row. The spinner is particularly useful in wet soil conditions, but can cause damage to the tubers.

Spiv Ewe. A ewe in poor condition and which will not fatten.

Splice Grafting. Method of grafting in which oblique cuts are made across both scion and stock of almost equal diameter. The cut surfaces of scion and stock are placed together and are tied after matching their surfaces.

Spliced Approach Graft. Type of approach grafting wherein the choosen two stems are approximately of the same size. A slice of bark and wood 2 to 5 cm long is cut from both the stems, at the point where the union is to occur.

Spoil. Excavated material from ditches or rivers.

Spraing. A disease of potatoes caused either by tobacco rattle virus (T.R.V.) carried by a free-living nematode worm or by Potato Mop-Top Virus (P.M.T.V.) transmitted by the powdery scab fungus. It is characterised by rust-coloured arcs in the flesh of the tubers.

Spray. A liquid applied under pressure, via a nozzle, in the form a mist of fine droplets. Insecticides, herbicides, liquid fertilizers and disinfectants etc., are often applied as sprays.

Spray Drift. The tendency of a spray of fine droplets, produced by a low volume nozzle, to drift in the air from the field of application to other fields. Such drifts may cause damage to other crops and can poison grazing animals or bees if the spray contains harmful chemicals. Drift damage is minimised by spraying in suitable weather conditions and by using sprays of sufficient dilution. The addition of a chemical 'drift retardant' to a spray mix may reduce spray drift by reducing the number of small drift-susceptible droplets. The chemicals act by increasing the viscosity of the spray mix.

Spray Irrigation. The irrigation of crops (*a*) by pumping water along distribution pipes to various types of distributor (*e.g.*, large nozzle rain guns, small nozzle rotary sprinklers, fixed spray lines), or (*b*) by applying water from a tank carried on either a water-operated vehicle fitted with a spray boom or oscillating rain gun, or on a tractor or other vehicle fitted with a sprinkler boom.

Spray Race. A type of race along which nozzles and fitted at intervals. Used to treat cattle and sheep with liquid parasiticides.

Sprayer. A machine used to apply a spray by forcing the liquid through a nozzle under pressure. Field crop sprayers are classified in terms of their capacity for application, as follows:

Volume	*litres/ha*	*gal.ac.*
Low	less than 220	less than 20
Medium	220-660	20-60
High	more than 660	more than 60

Sprayers essentially consist of a tank from which the liquid is pumped to the nozzles which are fitted at regular intervals along a boom. Nozzles may be of the fan type (used for low and medium volume spraying) or cone type (used for high volume spraying). Sprayers with tanks up to 450 1 (*c.* 100 gal.) capacity may be tractor mounted. Those of larger capacity, up to 1800 1 (400 gal.), are either trailed or mounted on 4-wheel drive vehicles.

Common types of spray nozzle produce a wide range of droplet sizes. In recent years sprayers have been developed which use high-speed of spinning nylon discs with fine-toothed margins to produce droplets of uniform size in the range 250-300 microns. These sprayers use less spray, give good crop cover and produce less spray drift. Other sprayers designed to apply small volumes of liquid (generally less than 5 litres/ha) (Ultra Low Volume-U.L.V.) use a droplet size of 60-70 microns which rely on being carried in the wind to give thorough crop cover. U.L.V. sprayers are prone to spray drift and are not used to apply herbicides or poisonous chemicals. Electrostatic U.L.V. sprayers are now being developed. These produce positively charged droplets of controlled size (*c.* 50 microns) which behave fairly predictably and are attracted to the mainly uncharged target crop. Spraying may also be carried out by aeroplane or helicopter.

Sprig. 1. A shoot or twig. 2. A Scion 3. A tiny spike sometimes present on the side of a single comb in poultry. Also called side-sprig.

Spring Cleaning. The ploughing of a field in the spring as th soil dries, following ploughing during the winter, to speed the drying process and to rise weeds (*e.g.*, Cough Grass) to the surface. Further cultivation and harrowing reduces the soil to a tilth allowing the weeds to be removed, usually by chain harrow, heaped and burnt. If they are fully dry they may alternatively be ploughed in. Usually carried out prior to sowing late-planted rootcrops.

Spring Cultivations. Those cultivations carried out on the land in the spring when the frosts are over, in preparing a seed bed, e.g., rolling, discing, harrowing etc.

Springer. A cow almost ready to give birth to a calf. Also called a down-calver.

Spring-tined Harrow. A versatile cultivation implement, the tines of which are sickle-shaped and made of spring steel, so that they are able to vibrate and shatter the soil. The angle of the tines may be varied giving different working depths suited to the type of cultivation required, so that the implement can be used either as a light harrow or a light cultivator.

Spud. A Common term for a potato tuber.

Spur. 1. A horny, claw-like growth on the back of the leg of a cock or other bird. 2. A short side branch on a tree or other woody plant, usually bearing flowers or fruit. Also a lateral root.

A hollow pouch-like projection at the base of a petal, often containing nectar.

Square Ploughing. A method of round and round ploughing, sometimes used for deep work in fields of more or less regular shape. A land with a similar configuration to the field boundaries is ploughed in the centre of the field. The field is then ploughed in a clockwise manner round this land. The plough is raised out of the soil at corners. This method obviates the need for open furrows.

Squeaker. A piglet.

Stable Fly. A greyish fly which breeds in faeces and rotting vegetation. They are a serious pest of horses, cattle and pigs in the warm summer months when they bite their legs to obtain blood, causing severe irritation and restlessness. They are often found in buildings housing livestock, *e.g.*, stables, cowsheds, etc.

Stack. A large pile of hay, straw, or corn.

Stackyard. An enclosed area or farmyard in which a stack of hay or straw is built. Also called a rickyard, and in the case of hay a hayguard.

Staddle. 1. A support structure on which stacks are built. Also called steddall or steddle. 2. A small tree left unfelled. Also a stump left to produce coppice. 3. A rootstock tall enough to produce a standard fruit tree when a scion has been grafted on.

Stag. 1. An adult male deer or male trukey. 2. An animal castrated when mature, particularly a boar (also called brawner, seg or steg).

Stage Set. A set or young hop plant which has remained in a nursery bed for more than one year.

Staggers. A condition of animals in which they are caused to stagger about.

Stale Furrow. Land which has been left for some time after ploughing so that the soil has had time to settle down and consolidate.

Stall. A partitioned compartment in a cowshed or stable in which cattle or horses stand or lie down.

Stallion. An uncastrated male horse, of 5 years of age or more. Mainly kept for breeding. Also called an entire.

Standard. 1. A fruit tree, the lower branches of which are about 1.8m (6 ft) above the ground. A half-standard is one in which the branches are about 1.2m (4 ft) above the ground. 2. A tree left to grow in coppice woodland. (Coppice with Standard). 3. An upright post supporting fencing wire.

Standard Crop. A crop growing in a field.

Standard Furrows. A term for those furrows still unploughed. Also called green furrows.

Standings. The raised parts of the floor of a cowhouse on which the cows are kept in pairs in stalls, each normally about 1.5m (5ft) deep and 2.1m (7 ft) wide.

Staple. Wool fibres naturally binding together in a fleece forming a lock. The staple length of wool is an indication of the average length of the wool fibres.

Starch. A polysaccharide carbohydrate, one of the main energy storage substances of plants, formed as a product of photosynthesis. It is stored in granule form in seeds (particularly cereal grains), tubers (*e.g.,* potato) and roots, which are consequently important in the diet of livestock and humans as sources of carbohydrate. Starch occurs in two component forms, amylose and amylopectin. Cereal and potato starches contain 20-30% amylose and 70-80% amylopectin. On digestion starch is broken down, first to maltose and then to glucose. Carbohydrate is not stored in animals in the form of starch but as glycogen, sometimes called animal starch.

Starch Equivalent. A term used in calculating livestock rations to indicate the feeding value of feedingstuff. The S.E. system, developed by the German nutritionist Kellener, gives values expressed as percentages which indicate the number of libs of pure starch having the same energy value as 100 lbs of a particular feedingstuff. S.E. values vary considerably, *e.g.,* Barley 71%, Soyabean Meal 64%, Kale about 9%, whilst grass silage can vary from 0.1% to 12%. The use of S.E. values is gradually being replaced by metric Metabolisable Energy values.

Steading. Farm buildings, with or without a farmhouse.

Steamed Bone Meal, Bone Flour. Steamed bone meal is an organic phosphatic fertilizer consisting of bones which have been gently steamed to remove fats and gelating for use in glue manufacturing. Steamed bone flour is the product of more intensive steaming followed by grinding the bones.

Steaming Up. The practice of feeding concentrates in addition to normal maintenance rations to pregnant cows or heifers during the two months prior to calving, in to build up their condition for the commencement of lactation.

Stecklings. Young Mangel or Sugar Beet plants grown in a seed bed for later transplanting to produce seed crops, or alternatively undersown in barley and allowed to develop into a seed crop *in situ.*

Steer. A castrated male ox over 1, year old. Also alled a bullock or stott.

Steerage Hoe. A type of Hoe, rear-mounted on a tractor, requiring an additional operator to the tractor driver, who is able to steer or guide the implement to avoid crop damage. Steerage is effected either by a direct linkage mechanism from the tractor or by controlling the depth of ground wheels.

Stell. An open stone or corrugated metal enclosure or shelter for sheep and cattle, usually circular, built hills and moorland.

Steam Eelworm. An eelworm which is a pest of oats and red clover, causing seedings to become distroyed and swollen, and in oats causing the condition known as 'tulip root'.

Sterile. 1. Infertile or barren, and unable to produce offspring, fruit, seeds, spores, etc. 2. A term for land unable to grow crops. 3. Lacking micro-organisms, particularly those causing disease, *e.g.,* sterile surgical instruments.

Sterilization. Destruction or removal of all living bacteria, micro-organisms, and spores. It is accomplished either physically (by heat or other radiant energy and filtration) or chemically (by germicides).

Sterilised Milk. One of the types of liquid milk sold for public consumption. Such milk must pass the turbidity test after prolonged heating, sometimes for up to 30 minutes, to between 110° and 115°C. (230°-240°F). This heating, carried out in sealed airtight containers in which the milk is later sold, kills all the bacteria present and further contamination is prevented by the airtight conditions. Such milk has a storage life of at least 7 days but the sterilising process affects its flavour and colour.

Stetch. A type of ridge, about 2.4m (8 ft) wide, separated from the next by an open furrow. Such ridges and furrows have been ploughed in low-laying parts of suffolk and essex for many generations to remove surface water.

Stibbler. A horse put out to graze on Stubble.

Stibestrol. Drug used in cattle feeding which stimulates gain and feed efficiency with estrogenic-like activity.

Sticking. A term used at abattoirs for bleeding livestock, by cutting their throats, after they have been stunned and hoisted.

Stilts. The handles of a horse-drawn plough.

Stimulant Crops. Crops whose usable part of the plant stimulate the nervous system *e.g.*, tea, coffee, tobacco etc.

Stock. 1. The main stem of a plant, particularly the trunk of a tree. 2. Graft. 3. The perennial parts of herbaceous plants. 4. The race, family type or source material from which a plant or animal has been bred. 5. The animals on a farm (livestock) or the various stores, implements and other equipment (deadstock). Also to purchase such livestock and deadstock for a farm. 6. To sow a ley. 7. To graze animals in a field. 8. A colony of bees. 9. To impregnate a female animal. 10. A hill.

Stock Nitrogen. The nitrogen present in animal faeces as distinct form in artificial fertilizers.

Stock Seed. Seed supplied by a merchant to a seed grower from which further supplies of seed may be grown.

Stocking. Distinctive coloured hair or feather on an animal's leg.

Stocking Rate. The number of grazing animals that are allowed to graze a given area of pasture, normally expressed in terms of livestock units per acre or hectare.

Stockman. A man in charge of livestock.

Stook. A number of sheaves (Sheaf) of corn set up in a field in 'pyramid' form to support each other, whilst drying out. Also called hile, shock, stitch or trave.

Stool. A tree stump from which shoots (stool shoots) sprout. (Stooling).

Stooling. A method of propagation in which the rootsocks of apple and pear trees are cut off at ground level to produce a stool from which shoots sprout. These stool shoots, which produce their own roots, are cut off and planted to produce new rootstocks.

Storage Drying. A method of drying bales of hay in a barn over a ventilated floor using a flow of unheated air. Initially about four layers are dried, and subsequently further layers are added and dried until the barn is filled. The bales are then stored *in situ.* Also called deep storage drying and barn hay drying.

Store. An animal kept at a steady rate of growth prior to later fattening for market.

Stover. A term for various types of fodder such as clover hay, cereal stubble and the broken pieces of straw from threshing.

Straight. An animal feed composed of only one type of feedingstuff, which may or may not have been processed before sale. (Compound Feed).

Straight Fertilizer. A fertilizer containing only one substances, usually providing only one, but sometimes two, of the major plant nutrients (*i.e.,* nitrogen, phosphorus and potassium). (Compound Fertilizer).

Strain. A breed, race stock or type within a species of plants or animals, the individuals of which have specific characteristics distinguishing them from those of other strains.

Straw. 1. A term mainly used for the dry stalks of cereals but sometimes applied to the Haulm of peas and beans. It is used for litter, thatching and as feedingstuff. It is less digestible and has a lower feeding value than hay, having a high Fibre content and a lower content of protein, minerals and vitamins. It is usually fed in the long form but sometimes chopped or ground, with other feedingstuffs such a concentrates. Wheat and rye straws have the lowest digestibility and energy value and are best suited for litter. Straw is mainly stored in bale form. 2. A container holding one dose of processed semen for use in artificial Insemination.

Straw Spreader. A mechanism (often a rotating disc) sometimes fitted behind a combine harvester at the straw outlet to spread the straw on the field.

Straw Yard. An enclosed area covered with straw as litter, in which cattle are kept during winter.

Stretch. A set of 5 or 6 hurdles together with stakes and shackles.

Strike. 1. The infestation of the flesh of sheep by maggots hatched from the eggs of the blow fly laid in the fleece. This condition, in which sheep are said to have been 'struck', causes intense irritation and death can occur within a week. Protection is afforded by Dipping. 2. To take root.

Strip. To remove the last few drops of milk from a cow's udder.

Strip Cup. A small container into which the first few squirts of milk are drawn by hand from a cow's teats before milking starts. This foremilk is discarded as it has a high bacterial content.

Strip Grazing. A grazing system whereby cattle are allowed access to a limited area of fresh pasture up to twice daily by means of a movable electric fence. As each strip is grazed a 'back fence' is also moved forwards to protect the grazed area to allow it to recover. This is an efficient method which provides for accurate rationing, and ensures that the cattle eat the entire plant (*i.e.*, stems as well as leaves) and limits damage by trampling and fouling. When grazed down the field is allowed to rest for 3-5 weeks before being grazed again.

Strippings. The last drops of milk drawn from a cow's udder during milking.

Stroke. To till a field once with a harrow.

Strong Land. Land covered with a heavy soil.

Strong Pasture. Grassland which is too rich (*i.e.*, has a high protein content) for young animals to graze, *e.g.*, the Early bite.

Strong Wheat. Varieties of wheat, the grain of which produces good quality backing flour, rich in gluten. The grains are of a steely appearance as distinct from the white, mealy grains of the inferior quality 'weak' wheats which are usually used for biscuit marking or as livestock feed.

Stubble. The part of a crop left in the ground after harvesting, including the roots and short stem remains, particularly the dry stalks of cereals. Also called gratten.

Stubble Cleaning. The cultivation of the land in the autumn after harvest, either with a Plough or heavy Cultivator, followed by harrowing to form a tilth and free the weeds from the soil so that they may be removed, usually by a chain harrow, heaped and burnt. Further cultivating destroys any seedlings which germinate. Also called autumn cleaning.

Stubble Mulch. Minimum tillage farming practice in which weeds and volunteer grain are killed but the stubble is left standing to catch drifting snow and facilitate penetration of water into the soil thus reducing erosion and improving moisture condition of the soil.

Sturdy. A disease of sheep and sometimes cattle, characterised by dizziness or staggering, caused by tapeworm cysts (also called blebs) in the brain membranes. The cysts are formed by embryos of the dog tapeworm, *Taenia caenurus,* hatched from eggs deposited in grass in the faeces of infected dogs and subsequently eaten during grazing. Also called dunt, gid, gig, goggles, turn, turnsick, thorter-ill and pendro.

Subsoil. The layer of soil below the topsoil, of lower organic matter content.

Subsoiler. A type of heavy cultivator with long tines or 'chisels' drawn through the soil at between 30 and 60 cm (2 and 24 in.), and used to break up a compact subsoil without inverting it. Deep cracks and fissures are caused, improving drainage, air and root penetration, and soil structure.

Succession. 1. A rotation of crops. 2. The sequence in which plants colonise a new habitat or bare ground, in which each plant community is superseded until a climax vegetation develops.

Successional Cropping. 1. The growing of successive but different crops in the same field in the same season. 2. The planting of a crop in small areas over a period so that a little ripens at a time, ensuring a gradual continuous supply over a long period.

Succulent Foods. Those feedingstuffs with a relatively high water content, approximately 70-90%. They are highly digestive but are low in oil, protein and fibre. They are valued particularly for winter feeding to balance dry fodders which tend to cause constipation, and especially for pregnant, sick or high yielding animals. Examples include the carbohydrate-containing tubers and root crops such as potatoes, carrots, mangels, swedes, and turnips etc. Various by-products with a low dietetic value are sometimes regarded as succulent foods, such as wet brewers' grains, wet sugar beet pulp, liquid skim milk, etc.

Sucking Pig. A young milk-fed piglet.

Suckler Cow. A cow which is allowed to rear its own calf and is then used for beef production, as distinct from being used as a dairy cow for milk production.

Sucklers. The flowering hands of clover.

Sugar. A general term for any of the sweet, soluble, monosaccharide or disaccharide carbohydrates, *e.g.,* Glucose Fructose, Sucrose (Cane Sugar), etc. It is commonly applied to the latter, common table sugar, which is obtained from sugarcane and Sugar beet.

Sugar Beet. A rootcrop of the genus *Beta* which also includes fodder beet and Mangel, grown for its high content of sucrose. The sugar content varies, subject to the variety grown, soil type, seasonal characteristic etc., but average about 16%.

Sugar Beet Harvester. A machine (either tractor-drawn and p.t.o. driven, or self-propelled) used to harvest sugar beet, which cuts off the beet tops (crowns and leaves), lifts the roots from the ground, cleans off adhering soil, an deposits them onto an elevator mechanism which carries them into an adjacent trailer or storage tank. The majority of beet is now harvested by a complete harvester which carries out the above operations in one unit. Some two-stage harvesters are still used which have one topping unit and a separate lifting and cleaning unit.

Sulfiting. Practice of treating certain fruits and vegetable prior to dehydration, with low concentrations of sulphur dioxide or a salt of sulphurous acid dissolved in water to inactivate enzymes that promote oxidative darkening.

Sulphate of Ammonia. [$NH_4)_2SIO_4$] Ammonium sulphate, a nitrogenous fertilizer occurring as a soluble salt in the form of white needle-like crystais once produced as a by-product of coal-gas manufacture but now synthesised. In common with other non-metallic ammonium salts, when applied to fields the sulphate radical combines with calcium in the soil and is leached out, depleting the soil of lime and increasing acidity. It contains about 21% nitrogen.

Sulphate Potash. (K_2SO_4) Potassium sulphate, a fertilizer containing about 50% potash, occurring as a white crystalline powder, produced by the action of sulphuric acid on muriate of potash. Mainly used by fruit growers and in market gardens, and also on potatoes which produce tubers with a higher dry matter contents than when muriate of potash is applied.

Sulphur (S). A non-metallic chemical element, occurring as pale-yellow crystals in several forms, required by living organisms a constituent of protein (particularly keratin in animals) and some oils.

Sulphuric Acid. An extremely corrosive, colourless, oily liquid, which chars organic matter and is sometimes used as a Contact herbicide, *e.g.*, to burn off diseased potato haulm. Also called oil of vitriol.

Summer Feeding. 1. The supplying of water to marshland in the summer, usually by impounding river water to prevent natural drainage and the lowering of the water table. 2. The feeding regime of cattle in the summer months involving outdoor grazing with or without supplementary rations.

Sunflower. A composite plant the seeds of which are rich in edible oil, used for margarine, cooking oil and in medicine. The residue, after oil extraction, is used to produce cake or meal with a high Fibre content, sometimes used in Concentrates for cattle and sheep but not fed to pigs. The seed is sometimes fed to poultry, particularly during moulting.

Superphosphate. A type of fertilizer once widely used but now largely superseded by the more concentrated ammonium phosphate and triple superphosphate. It is a greyish, granular or powdery substance, produced by treating finely ground rock phosphate with a restricted amount of sulphuric acid, yielding mono-calcium phosphate, calcium sulphate and some unchanged rock phosphate. Superphosphate contains 18-21% water soluble phosphate.

Supplementary Rations. The concentrates fed to livestock in addition to bulk foods such as hay straw and roots.

Support Crops. Certain fast growing crops work as supports to vine crops *e.g.,* castor, shevri etc.

Surface Cultivations. Those tillage operations which are of a shallow nature and only affect the top soil, *e.g.,* to kill weeds, as distinct from those which involve deeper ploughing.

Surface Drainage. The practice of removing surplus water from land by surface channels, *e.g.,* ridges and furrows, as distinct from via subsoil drains.

Surface Water. Water unable to penetrate and drain through the soil and which finds its way to drainage channels, streams and rivers via the surface of the soil.

Sward. The carpet of grasses, clovers and other grassland species covering the ground in a pasture.

Swath. A band of grass, corn or other crop cut by a mower or scythe, lying on the ground.

Swath Turner. A tractor-drawn implement used in haymaking which gently inverts a swath exposing the underside to enable it to dry.

Swayback. 1. A brain disease of lambs due to an inadequacy of copper in the ewe's diet, which causes paralysis and is often fatal. Lambs are seen to stagger or unable to walk. 2. A horse with a sagging back.

Swede. A rootcrop (*Brassica rutabaga*) similar to the turnip, characterised by smooth, ashy-grey leaves growing from an extended stem or 'neck', marked by scars. By contrast, turnips have hairy, grass-green leaves arising direct from the bulb itself. Swedes have a higher nutritional value than turnips, with Dry Matter content of 10-13%. They also haxe a longer growing period, are more hardy and are easily stored for winter feeding to livestock. They are sometimes pulped and mechanically fed to mangers. They are also sometimes grazed off in the field by folding. White and yellow-fleshed types exist, and of the latter, purple, bronze and green-skinned varieties are available. They are often grown as a root break following cereals.

Sweet Corn. Varieties of maize, the cobs of which are sweet and harvested for human consumption.

Sweet-Cure. A cure for bacon increasingly used as a replacement for the wiltshire cure, involving the injection of brine as distinct from soaking in brine.

Sweet Soil. A fertile alkaline soil, as distinct from a sour or infertile acid soil.

Swill. Waste food from kitchens used for feeding to pigs.

Swine Fever. An infectious notifiable disease of pigs caused by a virus. It is characterised by fever, refusal to eat, foul-smelling diarrhoea, discharges from the eyes, distressed breathing and general weakness. The disease may be acute, particularly in young pigs, and death may occur in a few days or it may take a chronic form, mainly in older pigs, which remain ill for a long period and lose condition, but may not die. Also called hog cholera and pig typhoid.

Swine Paratyphoid. Infectious bacterial disease occurring in acute or chronic from characterized by fever, anorexia and red or purple discolouration of skin on certain parts of the body followed by diarrhoea. It is caused by *Solmonella choleraesuts* which is specially adapted to swine.

Swine Vesicular Disease. A notifiable disease of pigs, caused by a virus. The symptoms are identical to those of foot and mouth disease and it is only distinguishable by virological testing in the laboratory. Its spread is associated with the feeding of swill. It is easily transmitted by untreated animal waste, and swill must, by law, now be boiled before feeding. S.V.D. is controlled by compulsory slaughter of infected pigs.

Swing Plough. An old type of plough which lacked wheels and had a shortish mouldboard, so that it produced a greater digging action than ordinary ploughs.

Switch. 1. A long, flexible twig or shoot. 2. To prune or trim a hedge, tree, etc.

Symbiosis. The phenomenon of two different organisms living together for mutual benefit, *e.g.*, the association of nitrogen-fixing bacteria with leguminous plants in the root nodules. The bacteria obtain carbohydrates and other foods from the plants and fix atmospheric nitrogen (Nitrogen Fixation) which is made available to the plants in various compounds. Lichens consist of fungi and algae living together symbiotically.

Syrup Feeds. By-products of the whisky distilling industry which are turned into animal feeds, particularly for dairy cattle. They comprise a blend of malt or maize draff and distiller's syrups and have been shown to contain high sources of energy and protein.

Systematic Ploughing. The ploughing of field in lands.

Systemic Compound. A chemical which, when applied to foliage or the soil, is absorbed by a plant and moved in the sap to all parts of the plant. Insects sucking the sap of plants treated with systemic insecticides are poisoned. Translocated-herbicides are systemic compounds.

System of Farming. The combination of products on a given farm and the methods or practices that are used in the production of these products is known as system of farming.

T

2, 4, 5-T. An abbreviation for 2,4, 5-trichlor phenoxyacetic acid, the active ingredient in herbicide products used in forestry and other applications to control woody weeds such as brambles. All products containing 2, 4, 5-T must be used strictly in accordance with recommended precautions.

Tabanid. A fiy of the genus *Tabanus.*

Table Birds. Poultry reared for their meat as distinct from for egg production.

Tack. 1. Pasture rented for grazing livestock. 2. Harness equipment.

Tail Corn. Small, undersized or very light grains of corn in a sample. Also called tailings.

Tailpiece. An extension sometimes attached to the rear end of a mouldboard. It can be set to push the turning furrow further over.

Take. 1. The taking possession of buildings and land by a farmer when the assumes a tenancy. Also called entry. 2. A crop which germinates or a successful graft are said to take.

Take-all. A disease of cereals due to a soil-borne fungus (*Gaumannomyces graminis*) which infects the roots and stem base, causing dicoloration, stunted growth, premature ripening and the production of bleached ears containing little or no grain. One strain of the disease affects only wheat and barley, whilst another also affects oats but is mainly confined to the wetter areas in the North and West. Control is by rotation. Also called whiteheads.

Tank Mixing. A relative new practice of mixing crop-protection chemicals (and also liquid fertilizers, adjuvant oils, trace elements and foliar feeds) before spraying on field crops, particularly cereals. The benefits include savings in time, labour, water and fuel, reduced wear and tear on machinery and less crop damage through wheelings. Manufacturers recommended mixes are cleared and approved under the pesticides safety precautions scheme and the Agricultural Chemicals Approvals Scheme.

Tankard Mangel. A type of mangel which is tankard-shaped and classified with the intermediate types which have less of their swollen root above ground than the globe types.

Tapeworms. A class of parasitic flatworms which inhabit the small intestine of animals and are named from their dirty white tape-like appearance. They consist of a ribbon-like chain of segments anchored at one end and trailing through the gut, reaching several feet in length in some cases. They seldom require treatment in livestock.

Tassel. 1. The male flower of the maize plant, responsible for pollen production. 2. A black hairy tuft on a turkey's neck.

Tear. The soil not cut on the coulter face during ploughing.

Teart Land. The rich pastures of Somerest on heavy soils containing an excess of molybdenum, renowned for causing scour in sheep and cattle grazing on them.

Teat Cup. One of the four tube-like devices forming the cluster of a milking machine. Each teat cup consists of an outer metallic shell lined with synthetic rubber, which fits over the teat and into which milk is drawn by the application of a partial vacuum.

Teat Dipping. A method of applying disinfectant to a cow's teats as a precaution against the spread of mastitis, involving the dipping of each teat into a cup containing the disinfectant. The three conventional designs of cup are (*a*) a straight cup, (*b*) a cup on top of a 'squeeze' bottle, and (*c*) a non-squeeze, two chamber model. The most effective cups are at least 7.5 cm (3 in.) deep, which ensure complete immersion. Disinfectant may also be applied by sprayers but this is regarded as less effective. The three main disinfectants used are iodophors, hypochlorite and chlorohexidenes, and they are applied after washing the udder, but prior to milking.

Ted. To toss swaths of new-mown grass in haymaking in order to expose more grass to the sun and air and quicken the drying process.

Tedder. A tractor-drawn haymaking machine, usually used soon after a grass crop has been cut to loosen and aerate the swath.

Temporary Grassland. Arable land sown to a ley for a limited period of years, as distinct from permanent grassland.

Temporary Pasture. Vegetation grazed for a short period only ; not more than one full year.

Tentiometer (Soil Moisture). Instrument used for measuring tension with which moisture is held in the soil. It consists of poorus clay cell filled with water which is placed in the soil. A pressure gauge is attached to the cell which registers the changing tensions as water moves through the cell towards the drier soil or into the cell from wet soil.

Teschen Disease. A notifiable disease of pigs due to a virus, resulting in an excited condition with fever and paralysis, and often death. It was first recognised in Czechoslovakia.

Testicle. The reproductive gland of male animals occuring in pairs, producing sperm. Also called testis (plural, tests).

Tetanus. A bacterial disease (*Clostridium tetani*) of livestock and man which enters the body via wounds, particularly from the soil. Muscles become stiff, spasms may occur, and the muscle which closes the jaw may become continously contracted. Death usually follows. Few animals recover even if treated. Also called lockjaw.

Tether. A rope, chain or halter used to confine or restrain an animal within certain limits, e.g., by tying it to a post in the ground.

Tetrazolium Test. A biochemical test in which seed viability is tested, not by germination, but by the red colour appearing when the seeds are soaked in 2, 3, 5, triphenyltetrazolium chloride (TTC).

Texel. A breed of sheep originating in Holland and widely distributed in Europe. It is renowed for its milk production, and also noted for its rate of growth and potential for meat production (particularly in France where a distinct meat type has been developed).

Thermal Death Point. That temperature at which an organism is killed after an exposure of ten minutes under certain specified conditions.

Thin. To remove some of the plants in a crop so that it is less crowded and those remaining have room to develop.

Thin Crop. A poorly developed crop with plants growing relatively sparsely.

Thin Land. Shallow soil.

Thinning of Fruits. Removal of a few young fruits from heavily bearing fruits with a view to provide greater advantage of space, light, water and food. It is usually practiced in papaya.

Thirds. A term once given to a category of wheat offal resulting from milling. All wheat offals, apart from bran, are now sold as weatings for use in feedingstuffs.

Three-point-Linkage. The three hydraulically operated arms fitted to the rear of a tractor on which implements are mounted.

Thresh. To separate the grain or seeds from crops (e.g., cereals, beans, peas, clovers, etc.) by beating.

Thresher. A threshing machine.

Threshing, Blower Loss. Percentage of grains blown by blower along with bhusa with respect to total grain input in the thresher by weight.

Threshing, Cylinder Loss. Percentage of unthreshed gains from all outlets with respect to total grain input in the thresher by weight.

Threshing Efficiency. Percentage of threshed grain from all outlets of the thresher with respect to total grain input by weight is known as threshing efficiency.

Threshing Machine. A machine once commonly used to thresh cereals and other seed crops. Most cereals crops are now harvested and threshed by combine harvester.

Threshing Refraction. Components in wheat which differ from sound grain is called refraction.

Threshing, Sieve Loss. Percentage of healthy grains, dropped from sieve with respect to healthy grains received from main outlet by weight.

Threshing, Visible Damage Loss. Percentage of broken or cracked grains from all outlets with respect to total grain input in the thresher by weight.

Thrifty. A term applied to animals which are developing well, gaining weight, producing milk or eggs well, etc., and generally thriving.

Thrips. An order of small, sap-sucking flies (Thysanoptera), characterised by narrow wings with fringes of long hairs. Very many species exist, some of which are pests of crops, e.g., Onion Thrips and Pea Thrips. Also called Thunder flies. (Thrips is both singular and plural.)

Throw. 1. To give birth to. 2. To sire.

Thrunter. A three-year old ewe. Also called thrinter.

Thrush. An inflammation of the frog of a horse's hoof, accompanied by the production of a foul-smelling discharge.

Tick. A sub-order of large blood-sucking insects (Acarina) of the class Arachnida, allied to the Mites. They are serious parasites of livestock, and when present in large numbers they cause 'worrying' and lack of thrift. The best known is the sheep tick responsible for transmitting Louping-Ill, redwater and Tick-Borne Fever. Ticks are controlled by dipping, washing or spraying with insecticides. Also called faggs.

Tick-borne Fever. A disease of cattle and sheep caused by the micro-organism *Rickettsia* (intermediate between bacteria and viruses) which infects the white blood cells and is transmitted by the sheep tick. In catatle it causes fever, reduced milk yield for a period and increased susceptibility to other diseases (e.g., Redwater). In sheep the fever is accompanied by listlessness and loss of appetite, and pregnant ewes may abort.

Tile Drain. A baked, extruded clay pipe used for underground drains in land drainage systems. Such pipes replaced the original bent or horse-shoe clay tiles (from which the term 'title-draining' was derived) in the midnineteenth century.

Till. To work land with implements in preparation for a crop, e.g., ploughing, cultivating, harrowing, manuring, sowing, etc.

Tillage. Land lying fallow or under a crop, as distinct from under a Ley.

Tiller. A side-shoot from the base of a plant near to the ground, e.g., from the bottom of the stalk or stem of cereals and grasses.

Tilth. The physical condition of the topsoil after tillage. A fine tilth consists of small and loose, crumbling soil particles or mould. In a coarse tilth comparatively large clods constitute most of the broken material.

Timothy. A loose to densely tufted perennial grass (*Phleum pratense*), with hairless, broad, rough bluish-green leaf blades and flowerheads in the form of spike-like, cylindrical panicles.

Tine. A spike or prong on a cultivator, harrow, rake or fork. Also called point.

Tithe. A term once applied to a tenth of the produce of land and stock on a farm given to the church, and later to a cash sum of approximately equal value. A tithe barn was one in which corn paid as tithe was stored.

Tonguted Approach Graft. Type of approach grafting wherein the two stems choosen are approximately of the same size. Two cuts are made to form a tongue on each of the two stems. The second cut is downward on the scion to provide a thin tongue on each piece. A closely fitting graft union can be obtained by interlocking these tongues tightly.

Tool Bar. A metal frame fixed usually to the rear of a tractor on which the cultivating tools (e.g., hoes) for inter-row work are mounted.

Top. 1. The tallest growing grasses in a sward. 2. To lightly mow grassland to encourage more leafy growth. 3. To cut off the crowns and leaves of sugar beet. (Sugar beet harvester)

Top Dressing. Fertilizer applied to a crop after it has emerged from the ground.

Top Fruit. Fruit with relatively firm flesh, as distinct from the soft flesh of soft fruit. Divided into hard or pome fruits (e.g., Apples, Pears, quinces, etc.) and stone fruits or drupes (e.g., cherries, plums. etc.)

Top Wire. One of the parallel wires at the top of the wirework in a Hop Garden to which string is attached, up which the Bines grow.

Tops. The unwanted portions of sugar Beet plants cut off during harvesting, comprising the crown of the root and the leaves, used for feeding to cattle and sheep.

Topsoil. The surface soil disturbed during cultivation and rich in organic matter.

Total Digestible Energy Value. A percentage value which is a summation of the digestive nutrients (i.e., digestible crude protein, digestible crude fibre, digestible oil x 2.3, and digestible nitrogen-free extract) in a feedingstuff, determined by a digestibility trial. T.D.N. values are used to asses the energy value of pig and poultry diets. They are slightly higher than D-values since the high energy value of fats (oils) is taken into account.

Total Herbicide. A non-selective herbicide used to kill all types of vegetation, applied prior to crop planting, also used on foot-paths, roads, waste land, etc.

Toxic. A term usually applied to a chemical capable of killing, injuring or impairing an organism.

Toxin. A toxic substance, especially one produced by bacteria infecting part of an animal's body.

Trace Element. A chemical element required by a plant or animal in very small quantities for its metabolism. Such elements are often essential for the functioning of enzymes, hormones and vitamins. They include iron, zinc, boron, manganese, cobalt, molybdenum and copper and a lack of these elements may cause a deficiency disease (e.g., Heart Rot in sugar beet and Pinning in sheep, due to insufficient boron and cobalt respectively).

Deficiencies can also be included by the antagonistic action of other element which render otherwise satisfactory amounts unavailable to crops or animals. Also called micronutrient.

Traces. The ropes, chains or straps attached to the collar of a draught animal (usually a trace horse) by which it pulls an implement or cart.

Tracklayer. A type of tractor propelled by many small wheels running on two parallel, closed, articulated tracks. Also called a caterpillar or crawler tractor.

Tractor. A vehicle which has now almost entirely replaced the horse for pulling implements and trailers. The modern tractor is also able to operate various implements attached to its three point linkage and front loader arms, and to provide power to trailed and mounted machinery via the p.t.o. and hydraulic drive system. Some stationary machinery (e.g., sawbenches, Crushing mills, etc.) are also powered by tractors by belt pully.

A wide range of tractors are available and are usually classified according to engine power as indicated below. Tractors have increased in size and power over the years so that models over 100 h.p. are much more common now than in the late 1160s. The 75 h.p. machine is now regarded as a 'common workhorse' on arable farms. Two-and four-wheel drive types exist, and the latter may have equal or unequal sized drive tyres.

(*a*) *Less than 10 h.p. (7.6 kw)* : Usually single-axle and hand operated. Used in market gardens for hoeing and cultivating.

(*b*) *10-40 h.p. (7.6-30 kw)* : Usually two wheel drive, manoeuverable, with good ground clearance, sometimes with narrow wheels. Used for row-crop work.

(*c*) *40-75 h.p. (30-57 kw)* : Mostly two-wheel drive but some four-wheel drive and tracklayers. The general-purpose tractor used for a wide range of jobs.

(*d*) *75-100 h.p.* (*57-114 kw*) : Large models, both two-and four-wheel drive, and tracklayers. Used on large farms and by contractors for heavy work.

(*e*) *More than 150 h.p.* (*114 kw.*) : Large four-wheel drive types and tracklayers.

Since 1970 new tractors have been required to be fitted with a Safety Cab or safety frame (unless exempted).

Tractor Hoe. A type of hoe consisting of a rigid or spring-loaded blades attached to the toolbar of a row-crop tractor. It incorporates a set of A-shaped blades which loosen the soil between the rows, behind which are fitted a set of L-shaped blades which cut the roots of weeds. Shallow-rooting weeds are disturbed and die on drying. Discs are sometimes mounted between the two sets of blades, set an angle, to prevent damage to crop seedlings by soil covering.

Tractor Vaporising Oil. A type of fuel at one time widely used in tractors, consisting of paraffin with added aromatic hydrocarbons to prevent preignition, enabling its use in the internal combustion engine. Almost all tractors now use diesel fuel although some old models are still run on t.v.o. Some farmers prepare t.v.o. on-farm by adding petrol to paraffin approximately in the ratio 1 : 9.

Traditional Farm-fresh Poultry. Poultry plucked but with the viscera, head and feet intact. Formerly referred to as New York dressed poultry.

Training. When a plant is tied, fastened, staked or supported over a trellis or pergola in certain fashion or some of its parts are pruned with a view to giving the plant a framework, the operation is called training e.g., training of grapes.

Tramlines. Accurately spaced pathways left in a growing crop or field area to provide wheel guide marks for a tractor driver or machine operator to follow during subsequent operations. Tramlines enable the accurate application of pesticides, fertilizers or other products, facilitate precision during cultivation and lighten the load or task of the operator. They can be produced by omitting to sow seed along particular 'paths' during drilling by crop removal after emergence by chemical or other means, and by continuous wheeling.

Translocated Herbicide. A herbicide with when absorbed by a plant via the leaves (following foliar spraying) or via the roots (after application to the soil) moves in the sap to all parts of the plant, eventually killing it, e.g., 2,4-D. Translocated herbicides are systemic compounds.

Transplanter. A tractor-drawn machine used mainly for transplanting Brassica crops. Various designs exist. In the simple hand-fed machine the operator is seated close to the ground and places the plants into a furrow opened by the planter. The soil is firmed around the roots by a pair of following, inclined press wheels. An assistant is sometimes required to prepare and hand plants to the operator from a supply tray fitted to the machine. In other types the plants may be conveyed to the furrow in fingers on a continuous conveyor belt, or in other 'feeding' devices.

Transplanting. Act of moving the plant from one place to another with the object of growing it permanently at the new location.

Transported Soil. A soil comprised of material carried from elsewhere (eg., by wind, water, ice, gravity, etc.) and deposited such as alluvium, colluvium, etc.

Trap Nest. A type of nesting box which a hen or duck can enter via a trap door but cannot leave until released. Such devices are used to record the number and size of eggs laid, to assist in selecting hens for breeding.

Tree Fruit. Fruit which grows on a tree (e.g, apples, pears, cherries, etc.) as distinct from those growing on bushes, canes or low-growing plants.

Trefoil. An annual leguminous plant (*Medicago lupulina*), with trifoliate leaves, small yellow flowers and twisted, black pods. Abundant on lime-rich soils. It is usefully for short-term leys and for catch crops and produces a early bite for sheep on chalk land.

Trenching. A method of deep digging in which adjacent trenches are dug two spits deep, with the top spit being turned into the bottom of the last (adjacent) trench dug, and the bottom spit being turned on top. Soil layers are reversed, as opposed to being replaced in their natural order as in bastard trenching.

Trench Layering. An asexual reproductive process of plant propagation involving laying down the whole stem. The new shoots are thus forced to push their way through a layer of soil which prevents the bark from colouring and favours quick root formation.

Tribunal. Agricultural land tribunal.

Trichinosis. A disease of pigs and other animals, including man, in which muscles become infected with the larvae of the small roundworm *Trichinella spiralis* Pigs are usually infected after eating infected raw swill.

Trichomonas. A genus of protozoa, *T. foetus* which infects the genital passages of cows, causing abortion, pus formation, and sometimes sterility.

Trifolium. The genus of clover. Particularly applied to Crimson Clover, *Trifolium incarnatum*, a tall, erect annual plant with crimson flowers. It is frost-sensitive and cultivated mainly as a catch crop. It is sown in lightly harrowed corn stubble and sheep are grazed on it in late April and May in the following year, early, medium and late strains are available, which can be grazed in sequence.

Tri-Iobobenzoic Acid (TIBA). Compound used to cause fruit set under otherwise adverse conditions.

Trotter. The food of a pig or sheep.

Trough Room. The amount of space required by each animal in a group feeding from a trough.

Truck Crops. Crops which yield in tons and grown for distant market requiring heavy transport.

Truck Gardening. Growing of crops like potato, onion and cabbage etc. on large scale for distant market.

True Protein. An assessment of the actual protein content of an animal feedingstuff derived by subtracting the non-protein nitrogen from the total nitrogen content. Roots and silage contain significant amount of non-protein nitrogen.

Truss. 1. A bundle of hay or straw. 2. A cluster of fruit or flowers, e.g., tomatoes. 3. To prepare a bird for cooking after killing and plucking, by removing the head and viscera, and tying or skewering.

Trusser. A mechanism sometimes fitted to the back of a combine harvester which trusses the straw leaving the combine into bundles. It avoids the need to subsequently use a baler on the straw deposited on the field. Also called a buncher or straw press.

T-Sums. A system, of dutch origin, which purports to indicate the best time to apply nitrogen to make the best use of grassland. It involves accumulating the daily average of maximum and minimum air temperature (°C) from 1st January until the total reaches 200, at which point application is carried out. Positive temperatures are added cumulatively whilst negative ones are ignored. The date at which T-Sum 200 is reached varies in different parts of the country.

Tuber. A swollen underground stem (stem tuber) bearing buds in the axils of rudimentary leaves or scales, e.g., potato. Also a swollen root (root tuber), e.g. dahlia. Tubers are organs of vegetative reproduction and contain stored food, e.g., in the form of starch in the potato.

Tuberculin. A liquid preparation derived from a culture of tubercle bacilli, containing tuberculo-protein, injected into the skin of animals of determine if animals are infected with tuberculosis (the Intradermal Comparative Test). Infected animals react by producing a swelling at the site of injection.

Tuberculosis. A contagious disease of many animals including man and all domesticated animals, particularly cattle and pigs, caused by the bacterium *Mycobacterium tuberculosis*. Characteristic tubercles or nodular swellings develop in one or more of the organs of the body and degenerate, becoming cheesy, so that affected organs are destroyed. T.B. of the lungs is the most common and is usually associated with coughing (chronic in cattle) and loss of condition

Tumbrill. A tip-cart used for carrying dung.

Tump. A hillock or mound. Also a clump.

Tunnel Drying. A method of drying bales of hay built into a stack with a central tunnel blocked at one end, so that air blown into it with a fan is forced up through the stack. The strack can be constructed in the field or more usually under a dutch barn when it is also called barn hay drying.

Turbidity Test. A statutory test which must pass before it can be sold for public consumption as sterilised milk. The principle of the test is that when milk is boiled all the albumen in it is precipitated. It involves the addition of ammonium sulphate to precipitate other substances, e.g, casein, which are then filtered out.

The filtrate is heated and any albumen in the milk is revealed by turbidity. If the milk had been adequately sterilised (by heating to in excess of 100°C) then the albumen would all have been previously precipitated, and the test would reveal no turbidity.

Turkey. A large poultry bird of the pheasant family native to North America several varieties of which are kept for their meat. Nowadays the main producers are specialists keeping large numbers of birds, although some small-scale producers still keep turkeys, themselves undertaking the killing plucking and dressing of ovenready birds (mainly at Christmas).

Turkey Grower. A turkey between 8 and 26 weeks of age.

Turkey Lion. A female turkey older than 26 weeks. (Turkey Grower).

Turkey Cock. An adult male turkey older than 26 weks. Also called turkey stag, or tom.

Turn. A term used of a female animal which comes on heat again after being mated. Also called return.

Turn Up. To lift a sheep so that it sits on its rump to facilitate shearing, veterinary treatment, etc.

Typhus. Acute contageous disease of animals which is characterized by high temperature and great prostration.

Twinter. A two-year old sheep or other animal.

Two-crop Ewe. A ewe that has borne two crops of lambs.

Two-tooth Sheep. Sheep of about 12-15 months of age.

U

Uberous. Yielding milk in abundance.

Udder. The organ containing the mammary glands of animals. It is generally applied to the pendulous bag hanging beneath a cow just in front of the rear legs, bearing four teats. Also called bag.

Udder clap. Mastitis.

Ultimate Wilting Point. Moisture content of the soil at which all the leaves of plants growing in that soil are completely wilted and will not recover in an approximately saturated atmosphere without addition of water to soil. Lower end of wilting range.

Ultra Heat Treated Milk. A designation of milk sold for liquid consumption after subjection to ultra high temperature to sterilise it by killing bacteria and other organisms, including spores, present in it. The milk is flow heated and held at a temperature of at least 270°F (132°C) for one second before cooling. This treatment destroys some vitamins and probably some proteins and afflects the palatability of the milk. Before sale U.H.T. milk must past the statutory Colony Count Test (to determine the possible presence of bacteria). It is sold in sterile packages, often cartons, to prevent recontamination, and can be stored indefinitely if kept sterile.

Unbroken Furrow. A continuous, smooth type of furrow slice such as that produced by a lea plough.

Undecorticated Seeds. Seeds which have not been subjected to decortication.

Underbeam Clearance. The measurement from the underside of the share to the underside of the beam of a plough.

Under Drain. A type of drain constructed beneath the ground, e.g. mole drain, tile drain, etc., in a land drainage system, as distinct from an open ditch or surface drain. The water draining into such pipes is carried away by gravity flow and is usually discharged into a ditch. Before the advent of tile drains the ducts were often formed from locally avaiable materials, e.g., thorn or other brushwood faggots, stones, dried peat or turf tiles, or straw bundles.

Undersow. To sow a grass seeds mixture with another crop (a Cover Crop), usually a cereal crop.

Undulant Fever. Refers to a recurrent feverish disease of man which can last for a long period, sometimes many months. It is caused by the bacteria called *Brucella abortus* which is responsible for Brucellosis and is usualy contracted by contact with infected cattle or by drinking unpasteurised infected milk. Also called human brucellosis.

Unexhausted Manurial Value. Refered to the amount of fertilizer, manure or lime remaining in the soil, and usefully available to a crop, after one or more crops have drawn upon it after its application.

Unfree Water. Water which is held closely by soil colloids so that it is not readily available to plants. It freezes well below 0°C.

Ungulates. The term used for the herbivorous hoofed animals, e.g., cattle, deer, horses, pigs and sheep.

Unit. 1. An implement which is directly mounted on a tractor as distinct from one trailed behind. 2. Livestock unit 3. Unit of account. 4. Unit of fertilizer.

Unit of Fertilizer. 1% of 1 cwt (i.e., 1.12lb). Thus a nitrogen fertilizer having 15% nitrogen in each cwt is said to have 15 units of nitrogen. A compound fertilizer having say 10, 15 and 20 units of nitrogen, phosphate and potash, respectively, contains 10%, 15% and 20% respectively of each per cwt (shown as 10 : 15 : 20). With metrication fertilizer application is increasingly expressed in kilograms per hectare.

Unkindly. A term applied (*a*) to soils which lack natural fertility or are cold, late, wet, etc., and (*b*) to crops or animals which lack thrift or are in poor condition.

Untreated Milk. Milk which is sold for liquid consumption direct from farms as opposed to being sent to dairies, and without any heat treatment (Heat Treated Milk).

Urban Fringe. The zone where the built-up area of a city, town or other urban area merges with the surrounding rural areas. Sometimes called the rural-urban interface or rurban fringe.

Urea $(CO(NH_2)_2.)$ A white, crystalline, organic compound which occurs in urine and was the first organic compound to be artificially synthesised. It is considered to be the most concentrated solid nitrogenous fertilizer, having 46% nitrogen, and is sold in a granulated form. It decomposes rapidly giving off ammonia which is converted to nitrate in the soil. Urea is very solube and is often used in liquid fertilizers and in concentrated compound fertilizers. It is also fed to stock as a source of non-protein nitrogen, mixed with a digestible carbohydrate source, e.g. Molasses, sugar-beet pulp, cereals, maize, silage, etc.

V

Vaccinate. Means to inoculate an animal with a preparation having dead or living, but weakened, antigens (e.g., bacteria or viruses) so that the animal produces antibodies in sufficient numbers to protect itself against the specific disease caused by the antigen concerned. The preparation is known as a vaccine.

Vaccine. Suspension of disease producing microorganisms modified by attenuation or by killing so that it will not cause disease and can stimulate the formation of antibodies upon inoculation i.e., vaccine is administered with the object of stimulating the recipient's specific defence mechanism in respect of given pathogen (s) or toxic agent (s).

Vacuum Silage. Silage prepared in a large, sealed, polythene 'bag'. The 'bag' is evacuated of air by a pump through a valve in the sheeting. This removal of air and the build up of carbon dioxide reduces respiration losses and aids fermentation, producing high quality silage.

Vapam (Sodium N-Methyl Dithiocarbamate Dihydrate). Water-soluble soil fumigant which kills weeds, germinating weed seeds, most soil fungi and, under the proper conditions nematodes. Vapam undergoes rapid decomposition to produce a very penetrating gas. The soil can be planted two weeks after its application.

Variegated. Of more than one colour, often in patches, e.g., the leaves of certain plants.

Veal. The meat from calves slaughtered at less than 15 weeks old.

Vegetable. A generally term applied to plants as distinct from animals. Normally applied to plants or parts of plants cultivated for human consumption or for stock-feeding, e.g., potatoes, carrots, cabbages, etc. Some fruits (e.g., tomatoes, cucumbers) and some seeds (e.g. peas, beans) are also considered as vegetables. Most vegetables are having useful amounts of vitamin C and minerals. Root vegetables have stored carbohydrates and seed vegetables are rich in protein.

Vegetable Dyes. Dyes prepared from various trees. They are known after the tree from which they are obtained e.g., santaline artocarpus and the cutch dyes prepared from the wood of *Pterocarpus Santalinus, Artocarpus integrifolia* and *Accacia catechu* respectively.

Vegetative Reproduction. Reproduction in plants not involving the flower or any sexual process, by the detachment of a part of the plant (e.g., rhizome, tuber, bulb, etc.), which then develops into a new plant. (Propagation)

Venison. Meat derived from deer.

Verandah. A type of housing which is used for poultry with a slatted or wiremesh floor raised above the ground, allowing the faeces to fall through. Most designs incorporate a sleeping area containing nest boxes and perches and a covered run in which troughs for food and water are located. The walls of the run are of wire mesh although the side walls may sometimes be partly boarded for protection.

Vermicide. A worm-killer agent.

Vernalization. Preplant treatment sometimes accorded to seeds in which they are placed under conditions favourable for germination for a short period and then are exposed to condition where the germination process is arrested.

Vetch. A leguminous plant *Vicia sativa*, with a slender, square, climbing stem, tendrilled pinnate leaves and blue or purple flowers. It is sown with rye to provide spring keep for sheep and with oats, or beans and oats, for silage. It may also be made into hay. Sometimes it is grown as a pure crop following cereals to produce seed. Winter and spring varieties are available. Also called tares.

Veterinary Science. Science and art of treating disease of animals. It encompasses anatomy, physiology, breeding, nutrition, animal diseases and treatment, man's use of animals, etc.

Viability of Seeds. It is represented by the germination percentage, which can be produced by a given number of seeds. A study of a viable seed is essential in successful propagation from seed. A reduction in seed viability may result from incomplete seed development on the plant, injuries during harvest, improper processing and storage or aging.

Veterinary Surgeon. A person qualified in veterinary science (the study of animal diseases), and able to diagnose diseases, prescribe medicines and carry out surgery.

Vineyard. A plantation of grapes.

Vining Peas. Peas grown for picking green, before they are fully ripe. For the past few seasons vining peas have been very unprofitable because rising production has coincide with a decline in demand from the processors.

Virgin Land. Land which has never been cultivated.

Virus. One of a group of sub-microscopic, self-reproducing, proteinous agents, which infect plants and animals causing disease (e.g., mosaic diseases, foot and mouth disease). They are transmitted between plants mainly by insects, particularly aphids, and by eelworms, and between animals by insects, contct and the inhalation of mucus droplets expelled by coughing and sneezing.

Virus Yellows. A mosaic virus disease of sugar beet or mangel plants due to a virus complex, *Beet Milk Yellow Virus* (B.M.Y.V.) and *Beet Yellow Virus* (B.Y.V.) spread by certain aphids The disease is characterised by the outer and middle leaves turning yellow from the tips and upper margins, and becoming thicker and brittle. Significant reductions in the sugar content and yield of crops are caused.

Viscera. The abdominal organs of animals, e.g., liver, kidneys spleen, etc.

Vitality of Seeds. Supply of vital seed is essential in successful propagation from seed. Vitality is represented by germination rate. It is often associated with viability of seeds. A reduction in seed vitality may result from incomplete seed development on the plant, injuries during harvest, improper processing and storage or aging.

Vitamins. A class of organic substances required by animals in small amounts, essential to metabolism and thought to act together with enzymes. Animals are able to synthesise several vitamin in their bodies but most must be supplied in their food. Vitamins A, D, E, and K are fat-soluble and are stored in the fat compounds of the body. Vitamins B and C are water soluble and cannot be stored and are regularly required in the diet. The known vitamins are listed in the table. *The Physiological Functions and Chief*

Sources of the Vitamins. All farm animals are able to synthesise sufficient vitamin C in their bodies. Ruminants and horses contain gut bacteria capable of producing sufficient supplies of the B group vitamins. Livestock kept outdoors are able during the summer to produce adequate supplies of vitamin D through the irradiation of their coats and skin.

The Physiological Functions and Chief Sources of the Vitamins.

Name	*Bodily functions for which required*	*Dietary sources*
Fat-Soluble		
Vitamin A	Vision, health of skin and walls of respiratory and urogenital tracts, growth and reproduction.	Some fish liver oils. Carotene (pro-vitamin A), fresh green leaves and carrots.
Vitamin D	Absorption and deposition of bone, growth and reproduction.	Some fish liver oils, irradiated yeast.
Vitamin E	Normal reproduction, Muscle health.	Wheatgerm, whole cereal grains, some vegetable oils.
Vitamin K	Clotting of blood.	Liver, egg, fresh green leaves.
Water-Soluble		
Vitamin B group B_1 (thiamine)	Carbohydrate metabolism, functioning of central nervous system.	Dried yeast, soya bean meal, liver heart.
B_2 (riboflavin)	Oxidation processesin the body	Dried yeast, dried milk, liver, heart.
Nocotinic acid	Oxidation processes in the body.	Dried yeast, dried milk, liver, heart, wheat and barley.
B_6 (pyridoxine)	Amino acid metabolism. Of greatest importance in meat eating species.	Cereals, dried yeast, dried milk, liver, egg yolk.
Pantothenic acid	Oxidation of fats.	Dried liver, dried yeast, dried milk, egg yolk, liver.
Biotin	Amino acid metabolism.	Wide distribution.
Choline	Fat formation, amino acid metabolism.	Most foods contain reasonable amounts.
B_{12} (cobalamin)	Amino acid metabolism. Normal growth, reproduction and blood formation.	Cheese, dried milk, liver, kidney.
Vitamin C	Normal digestion and carbohydrate metabolism.	Liver, green leaves, some fruits.

Deficiency in particular vitamins can cause ill-health and livestock diets usually include synthetic vitamin supplements. The latter are normally added to concerate rations in balanced amounts according to the animals' needs.

Viviparous. A descriptive term for animals which give birth to live offspring.

Volunteer. A plant which grows spontaneously in a crop of different plants, from seed present in the soil.

W

Walnut Comb. An additional fleshy comb on the heads of some breeds of poultry, resembling a half walnut, also called a strawberry comb.

Wapiti or Wapperty. A subspecies of deer originating from the U.S.A. First introduced into Britain at the turn of the century, there are very few now left.

Warble Fly. One of two species of fly, *Hypoderma bovis* and *H. lineata,* which cause great annoyance to cattle in the spring and summer when they lay their eggs on the underparts and legs. Cattle often rush about (known as gadding) to avoid attack, sometimes injuring themselves, and there may be a loss of condition and reduced milk yield. On hatching, the eggs burrow into the skin causing sores at the points of entry. The maggots migrate through the body and reach the back in January-June of the following year, where they cause swellings (warbles) about the size of a small walnut, sometimes pus-filled and each with a perforation or breathing hole. On maturity the maggots escape through the hide and fall to the ground to pupate. During migration in the body, the maggots cause damage to the tissues of the gullet and back, causing severe irritation and inflammation and loss of health and condition The value of the hides is also reduced due to the perforations. The condition is most serious in young animals and death sometimes occurs. Treatment is by the use of 'pour-on' insecticides applied to the back which are absorbed and kill the migrating maggots.

Warm Season Grass. Grasses that make their major growth during the warmer part of the year.

Wart Hot Disease. Highly fatal contagious disease of pigs similar to classical or European swine fever.

Warp. 1. To lay eggs or to give birth, especially prematurely. Also to miscarry, particularly in cattle and sheep. 2. Alluvial sediment.

Wraping. The natural build-up of a saltmarsh by the deposition of silt. 2. The controlled flooding of agricultural land adjoining a river with silt-charged river water so as to build up fertile soil.

Wart Disease. A notifiable disease of the potato, due to the fungus *Synchytrium endobioticum*, which causes the development of rough-surfaced, dark brown or black, tumorous growths on the tubers which, under favourable conditions, release motile spores which are able to infect young tubers, but which can persist indefinitely in the soil. Most varieties of potato grown now-a-days are immune

Waste Food. Defined as 'any meat, bones, blood, offal, or other part of the carcase of any livestock, or of any poultry, or product derived thereform, or hatchery waste or eggs or egg shells, or any broken or waste foodstuffs (including table or kitchen refuse, scarps or waste), which contain or have been in contact with any meat, bones, blood, offal, or with any other part of the carcase of any livestock or of any poultry.'

Water. Water is vital to all forms of farming. Livestock husbandry was found to be the largest consumer accounting for 35 000 million gallons (160 Mm^3), followed by irrigation using 20 000 million gallons (90 Mm^3) in a dry year. Average daily consumption of water by livestock is as follows.

	Daily Requirements
In pig sows	at least 4.55 litres
[illegible]	up to 22.75 litres
Fattening pigs	0.9-1.35 litres water per kg
Sheep	approx 4.55 litres
Fattening cattle	27-33 liters in water 33-36 litres in summer
Milking cows	4.55 litres per 50 kg body weight +4.55 litres per litre milk produced

For dairy cows water requirements are related to dry matter of ration in the following table.

Dry Matter of ration	*Daily milk yield (litres a cow)*				
%	10	15	20	25	30
	Water drunk (litres a cow)				
20	27	40	53	65	78
30	32	45	57	70	82
40	36	49	62	74	87
50	41	53	66	79	91
60	45	58	71	83	96
70	50	62	75	88	100
80	54	67	80	92	105
90	59	71	84	97	109

Water Table. The surface at which ground water has settled in the ground and below which fissures and pores in the rock strata or soil are saturated with water. This surface is uneven and its position varies according to the amount of rain which has fallen. Where a water table rises to intersect the ground surface, a spring arises.

Water-holding Capacity. A general term for the ability of a soil to retain water.

Waterlogged Soil. A soil which is saturated with water so that the pore space is completely filled with water, and conditions are unsuitable for plant growth since the roots are unable to obtain oxygen for respiration.

Watershed. The ridge of high ground separating the catchment areas of two distinct river systems.

Wattle. 1. The coloured fleshy skin beneath the throat of some birds Also the doddle of a goat. 2. A twig or flexible stick, woven with otherss to make a frame work for fences, roofs, hurdles, etc.

Waymark. To mark the course of a right of way, path or trail at points along it to enable users to use it accurately.

Weaner. A piglet which has been weaned (nowadays usually at 3-5 weeks) but has not reached the age of about 10 weeks when it becomes a feeding pig or 'fattener' for a variety of purposes.

Weathering. The process by which rocks disintegrate and decompose eventually producing soil particles, by exposure to the physical and chemical effects of atmospheric agents, e.g., rain water, frost, wind, temperature changes, plants and animals. Soils when formed, continue to degrade under the influence of such agents.

Weatings. A term now given to all wheat offals, other than bran, containing not more that 6% fibre, and used as a Feedingstuff. At one time such offals were grouped in three main categories; pollards, coarse middlings or sharps, and fine middlings or thirds.

Web Punching. The punching of holes in the webs between the toes of fowls and other poultry for identification purposes.

Wedge Shape. The conformation regarded as ideal for dairy cattle, the hindquarters being broader and deeper than for-equarters.

Weed. Generally herbaceous plant or shrub not valued for use or beauty, growing where unwanted, and regarded as using ground or hindering the growth of more desirable plants.

Weeding. Act of removing weeds from the infested area.

Weedkiller. A term for a herbicide.

Weedwipers. Devices which smear herbicides on weeds which protrude above crops. They are usually tractor mounted but hand applicators are available.

Weevils. A large family of small beetles (Rhynchopora) characterised by a beak-like prolongation of the head (the rostrum) and elbowed antennae, which cause damage, either as larvae or adults, to fruit, nuts, grain (e.g. the Grain Weevil, *Stiophilus granarius*) or trees. Pea and bean weevils (*Sitona sp.*) are unrelated (family, Bruchidae) and, as adults, cut notches in the leaves of the plants.

Well-sprung. A descriptive term for an animal with well arched ribs so that it is not flat-sided.

Went. A single journey across a field in cultivating, the return journey being known as a bout.

Westerwolds Ryegrass. An annual type of Italian Ryegrass (*Lolium multiflorum* var *Westerwoldicum*) which, when sown in spring or summer, rapidly sets seed and flowers. It is usefully used to provide hay in the year of seeding. Diploid and tetraploid varieties are available. Sometimes spelt Westerwolths.

Westrain. A cross-bred pig derived from a Belgian Pietrain and the old Wessex Saddleback.

Wet Feeding. The feeding to livestock of meal mixed with water in the form of a porridge.

Wet Fencing. Water-filled ditches and dykes in marsh areas which livestock will not or cannot cross, and which therefore act like fences in retaining animals.

Wet Mash. Meal fed to livestock after has been added to reduce it to a crumbly state.

Wether. A castrated adult male sheep. Also called wedder.

Wetting Agent. A substance which lowers the surface tension of water so that it spreads out rather than remains in droplet from when sprayed. Wetting agents are added to detergents to increase their efficiency and to sprays of insecticides, fungicides, herbicides or mineral sprays to improve the cover achieved when applied to leaf surfaces.

Whale Meat Meal. A feedingstuff with a high protein content, consisting mainly of Whale flesh, processed in a similar manner to Meat Meal.

Wheat. A cereal (*Triticum valgare*) grown for its grain, used mainly for flour making. Many varieties are available, the grains of which differ in size, shape and colour, and which exhibit different characteristics in respect of time of ripening, disease resistance, hardiness and adaptation to soil type and fertility. They fall into two groups, autumn and winter-sown varieties. The flowering head of wheat is a four-sided Spike bearing tightly-set spikelets, the palea of each ending in a point, and in some varieties (bearded wheats) extending into long bristly awn so that they resemble barley. Wheat is mainly grown in the drier eastern half of Britain, doing best on well-drained, fertile soils. It is usually grown following a ley or after a root or legume break in a rotation.

Wheat Bulb Fly. A two-winged fly (*Hylemyia coarctata*), the larvae of which bore into the central shoots of wheat in the spring (and also winter barley and ray) causing them to turn yellow, and die if tillers have not developed. Control is by seed Dressing using chlorofenvinphos and carbophenotion (Organophosphorous Compounds) and sometimes by spray treatment in the spring.

Whey. By-product of cheese making. For every 100 kg of milk used the yield of whey is 90. It contains most of the milk sugar and albumin of the original milk, calcium and vitamin B_2. The percentage composition of a typical whey is :

Water	:	93.40
Fat	:	0.35
Protein	:	0.85
Milk sugar	:	4.80
Ash	:	0.60

Whip Graft. A graft in which a tongue cut on the Scion is fitted into a slit cut slopingly in the Stock. Also called a tongue graft.

Whipple-tree Whippance. The crossbar of a horse-drawn implement or cart, pivoted in the middle, to which the Traces are fixed. Also called bullet, bodkin, swing or swingle tree.

White Clover. A perennial type of Clover (*Trifolium repens*), characterised by a creeping growth habit which knits the sward together, keeps out weeds and unproductive grasses, and minimises the destructive effects of poaching by livestock. Several varieties are available, all of which are persistent in the sward but are less bulky than Red Clover. White clovers tolerate grazing well and are used in all long Leys. (Dutch White Clover and Wild White Clover).

White Corpuscles. Leuocytes of the blood.

White Grub. Larva of various species of chafer.

White Mustard. A tall, branched, yellow-flowered, cruciferous plant (*Sinapsis alba*), producing long curved pods. It is grown as a Green Manure crop and sometimes for folding sheep on.

White Straw Crops. A term used for wheat, barley, oats and rye.

Whole. A term used to mean complete, e.g. (*a*) whole animals, which are entire and have not been castrated, (*b*) whole corn, which is unmilled, or (*c*) whole milk, which has not been separated. (Separated Milk).

Whole Milk. Milk which has neither been watered nor skimmed.

Whole Pressed. Having pressed the entire seed to remove oil.

Whole Work. Ploughing which leaves the furrow unbroken.

Wick. 1. A dialectal term sometimes used for a farm. 2. A term for a maggot.

Wild Oats. A tall, annual weed related to cultivated oats, but distinguished by having long, yellow or brown, hairy grains, each bearing a strong, twisted Awn. Wild oat infestations in cereal crops are a serious problem, increasing competition and reducing grain yield, and making harvesting, cleaning and drying more difficult and costly. They also lower the value of the crop.

Wild White Clover. A small-leaved, vigorous, truely perennial form of white clover, the stolons of which spread close to the ground keeping out weeds. It is drought resistant, persistent, and remains continuously palatable to grazing animals.

Wilting. A condition of plants due to loss of cell turgour as a result of water loss, characterised by the leaves and young stems drooping and becoming limp. Wilting can result from lack of water supply when transpiration exceeds root intake, and can also be induced by certain fungal diseases (wilt diseases). In haymaking and silage making, plants are often crimped or lacerated to increase wilting in order to reduce moisture content.

Wilting Point. The point at which the water content of a soil reaches such a level on drying out that it is all firmly held by the soil and is unavailable to plant roots, so that the plants wilt permanently and die.

Windbreak. A screen, particularly a line or clump of trees or a piece of woodland, providing protection from the wind to an area of land growing crops or livestock.

Windgall. A puffy swelling in the region of a horse's fetlock.

Windrow. A row of hay (or other cut crops), consisting of two or more Swaths combined into one in preparation for picking up.

Windrow Pick-up. A mechanism attached to the front of various harvesting machines (*e.g.,* Combine Harvester, pea harvester) to pick up a crop left in a windrow after previous cutting by a mower.

Wine Lake. A loose descriptive term for the E.E.C. structural surplus of wine, purchased as the Intervention Stock under the Common Agricultural Policy (Mountain).

Winnow. To separate the chaff from the grain of threshed corn by blowing it away in a draught from a fan.

Winter Feeding. The feeding regime of livestock, mainly cattle, in the winter months, involving the feeding of hay, silage, other bulk foods and concentrates.

Winter Green. A term applied to plants which remain green and fesh through thc winter.

Winter Wash. A spray applied to fruit trees and bushes during the winter months which soaks into crevices in the bark and kills the overwintering stages of pests and diseases. Usually A tar oil or similar organic compound.

Wirework. A term applied to the system of wires supported on a 'grid' of poles, erected in hop gardens, to which 'stringing wires' are attached, up which the hop bines grow. Various designs of wirework are in use.

Wireworms. The smooth, thin, yellow larvae of the click beetle which inhabit grassland and attack various crops, particularly cereals which are usually sown following a ley. Wireworm attack is greatest in spring and autumn when they eat into plants just below the soil surface, causing foliage to turn yellow before the plants die. Wireworms take 4-5 years to mature after which they pupate in the soil. Control is by the use of seed dressings containing B.H.C.

Witcher's Broom. Abnormal bushy growth of parts of the branch system on trees or shrubs, markedly different from that of the normal plant and characterized by the shortening of the internodes and excessive proliferation, generally pathogenic in origin.

Withers. The ridge between the shoulder blades of a horse, its highest part.

Wolf Tree. Tree occupying more space than its value warrants, curtailling better neighbourers. A term usually applied to broad-crowned, short-stemmed tree, or to a weed tree of large size; growing for some time, and overtopping others, or taking valuable space.

Wool. A soft, modified form of hair in which the fibres are shorter and curled, and which has an imbricated surface of minute, overlapping, interlocking scales which hold the wool fibres together. Wool is found on various mammals but the term is usually restricted to the fleeces of sheep. The annual wool yield per sheep varies from as low as 1 kg to 7 or 8 kg for certain long-woolled breeds.

Wool Ball. A mass of wool, tangled into ball, found in the first or fourth stomach of lambs. Wool is swallowed from the mother's fleece when suckling and accumulates, sometimes blocking the stomach outlet and causing death.

Wool Rot. A condition of sheep due to a fungus (*Dermatophilus dermatonomus)* which causes irritation to the skin in wet weather and the development of hard, yellowish scans. The sores heal and the growing wool carries the scabs away in the fleece. Also called Lumpy Wool.

Worms. A common name for a range of elongate, cylindrical legless invertebrates, many of which are parasitic.

Worrying. The tormenting or harassing of livestock, usually by uncontrolled dogs. This often involves the biting and tearing of flesh. Sheep, which become particularly anxious when worried,are frequently killed by dogs, and pregnant ewes often abort. Farmers are entitled to shoot dogs found worrying livestock.

Wry Tail.A term applied to poultry when the tail is carried turned to one side.

X

Xanthophyll. A carotenoid plant pigment (Carotene), yellowish in colour, present in chloroplasts and other plant parts where chlorophyll is absent. It assists in photosynthesis.

Xylem. Vascular tissue in plants which carries water and dissolved mineral salts from the roots, where they are taken in, to the various parts of the plant. It also provides mechanical support to plants, the cells being impregnated with lignin, and is located to the inside of the cambium. Secondary xylem comprises the bulk of the woody tissue in woody plants and is laid down by the process of secondary thickening.

Y

Yearling. An animal between 1 and 2 years of age.

Yeast.A single-celled ascomycete fungus, economically important in being capable of fermentation and an important source of vitamins.

Yeld. Dry or barren, not producing milk.

Yellow Rust. A fungal disease of barley and wheat, forming deep yellow pustules in closely parallel lines on the infected parts of leaves and stems. The disease commences on volunteer plants and spreads to both spring and winter crops, particularly when conditions are cool and moist. Spores are able to survive the winter.

Yelm. A bundle of straw used for thatching.

Yield of Crop. Produce obtained from a crop grown in a unit area (one acre or one hactre). It depens upon many factors such as type of soil, cultural practice, use of manures and fertilizers, timely irrigation, season and variety etc.

Yoke. 1. A frame joining a pair of draught oxen. Also any pair of animals working together. 2. A collar placed round an animal's neck to keep it in a stall.

Yolk. 1. The food store containing protein and fat granules in the eggs of most animals. The yellow central part of the eggs of brids. 2. The grease in wool. Also called euk, suint or yoke.

Yorkshire Fog. A perennial, soft, downy grass (*Holcus lanatus)*, common in poor pastures and an indicator of Lime deficiency in the soil. Also called meadow soft grass.

Z

Zeolite. A volcanic ash clay mineral, mined in rock form in the U.S.A. When groun and aplied to fields in a form similar to lime, it is claimed to have very strong ammonium ion holding powers which reduce the leaching of nitrates.

Zero Grazing. A grazing system in which grass (or other fodder) is cut daily and then taken to cattle kept in a yard or in a small exercise paddock near the buildings. The system obviates problems of poaching (particularly on heavy land), fouling, over-eating and wastage due to selective grazing. More grass can be provided from a given area by the system then by grazing. Also called mechanical grazing or soilage.

Zig-Zag Harrow. A type of Harrow in which the frames are zig-zag in shape, bearing short pointed tines, spaced so that each covers different ground. Used usually in the final stages of Seed Bed preparation, and also to remove tractor and drill wheel marks after sowing and to cover the seeds fully.

Zinc. Plant food element which is absorbed as zinc (Zn^{2+}) ion. It is considered essential in enzyme systems necessary for important reactions in the plant metabolism of life activity of plants. Zinc helps in the formation of some growth hormones, considered helpful in the reproduction of certain plants.

Zinc Arsenate. $Zn(AsO_2)_2$, a compound used as a component of some insecticidal sprays.

Zinc Chloride. $ZnCl_2$. A compound used as fungicide and wood preservative.

Zoonoses. A group of infectious disease transmissible from animals to man. The commonest include salmonellosis, leptospirosis, campylobacter infection and Q-fever.

Zootoxin. Poison or toxin produced by animals, as snake venom or spider poison.

Zooxanthin. Yellow pigment found in animals.

Zygote. A fertilised ovum before it starts to divide and form an Embryo.

Zymogenous Flora. Organisms found in the soil in large numbers following the addition of readily decomposable organic matter.